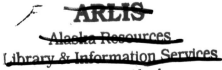

Arc Hydro

GIS for Water Resources

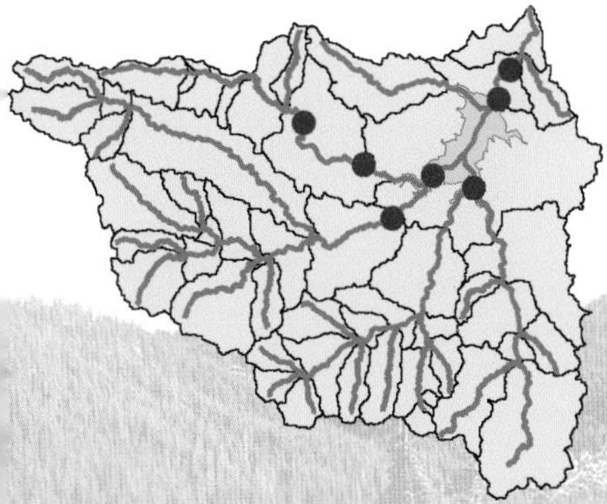

David R. Maidment, editor

ESRI Press

REDLANDS, CALIFORNIA

ESRI Press, 380 New York Street, Redlands, California 92373-8100

Copyright © 2002 ESRI

All rights reserved. First edition 2002
10 09 08 07 06 05   5 6 7 8 9 10

Printed in the United States of America

Library of Congress Cataloging-in-Publication Data
Arc Hydro : GIS for Water Resources / David R. Maidment, editor.
    p.   cm.
   Includes bibliographical references (p.   ).
   ISBN 1-58948-034-1 (pbk.)
   1. Arc hydro. 2. Hydrologic models. 3. Geographic information systems.
  I. Maidment, David R.
  GB656.2.H9 A735 2002
  553.7'0285—dc21   2002008865

ISBN-13: 978-1-58948-034-6
ISBN-10: 1-58948-034-1

Ask for ESRI Press titles at your local bookstore or order by calling 1-800-447-9778. You can also shop online at www.esri.com/esripress. Outside the United States, contact your local ESRI distributor.

ESRI Press titles are distributed to the trade by the following:
In North America, South America, Asia, and Australia:
Independent Publishers Group (IPG)
Telephone (United States): 1-800-888-4741
Telephone (international): 312-337-0747
E-mail: frontdesk@ipgbook.com

In the United Kingdom, Europe, and the Middle East:
Transatlantic Publishers Group Ltd.
Telephone: 44 20 8849 8013
Fax: 44 20 8849 5556
E-mail: transatlantic.publishers@regusnet.com

Cover design, book design and production, and image editing by Savitri Brant
Copyediting by Michael Hyatt and Tiffany Wilkerson
Printing coordination by Cliff Crabbe
Distribution coordination by Steve Hegle

Contents

Acknowledgments

I wish to acknowledge the generous collaboration and support that the Arc Hydro effort has received from the GIS and water resources communities. In particular, the following organizations have supported research efforts of the Consortium for GIS in Water Resources that was formed to guide the development of Arc Hydro: ESRI; U.S. Environmental Protection Agency; U.S. Geological Survey; U.S. Army Corps of Engineers; Texas Natural Resources Conservation Commission; Texas Water Development Board; Texas Department of Transportation; Lower Colorado River Authority; City of Austin, Texas; Dodson and Associates; Camp, Dresser and McKee; and DHI Water and Environment.

Individuals from many other organizations also contributed information and suggestions. In particular, I wish to acknowledge the contribution of the National Hydrography Dataset program of the USGS and EPA, whose development of a hydrography data model for the United States was a significant guide to this effort; the Topographic Science Program at the U.S. Geological Survey's EROS Data Center in Sioux Falls, South Dakota, which has been a beacon of progress for land-surface terrain modeling in hydrology; the Hydrologic Engineering Center of the U.S. Army Corps of Engineers, which has led the way in creating GIS tools for hydrology and hydraulics; the EPA Basins group, which has created a water quality data and modeling system that is the first comprehensive hydrologic information system; and DHI Water and Environment, which is pioneering the development of geospatial hydrologic modeling.

Other organizations who have made suggestions during the design of Arc Hydro include: Black & Veatch; Brown and Caldwell; CH2M Hill; Colorado Springs Utilities; Denver Water Board; Earth Tech Inc.; EMA Inc.; Haestad Methods; Horizon Systems Corporation; Johnson County, Kansas; MJ Harden Associates; New York Department of Environmental Protection; PBS & J; Storm Water Management Authority, Alabama; Tarrant Regional Water District; Tetra Tech; UMCES/Chesapeake Bay Program; University of Maryland; URS Consultants; U.S. Army

Engineer Research and Development Center; U.S. Forest Service; and Wisconsin Department of Natural Resources.

Many ongoing GIS in water resources research projects carried out at the Center for Research in Water Resources of the University of Texas at Austin created knowledge that was used in Arc Hydro. The following graduate students contributed to this effort: Kim Davis, Eric De Jonge, Kevin Donnelly, Jordan Furnans, Jon Goodall, Venkatesh Merwade, Katherine Osborne, Oscar Robayo, Victoria Samuels, Kristina Schneider, Sarah Stone, and Tim Whiteaker. In particular, I am grateful for the many contributions of Francisco Olivera, now of Texas A&M University. Chapter 5 in this book was superbly prepared by Nawajish Noman of ESRI and Jim Nelson of Brigham Young University.

Software designers at ESRI guided the Arc Hydro development effort, in particular, David Arctur, Dean Djokic, Steve Grise, and Dale Honeycutt. Scott Morehouse led this effort at ESRI, and Jack Dangermond was the inspiration for the formation of the Consortium for GIS in Water Resources. The Arc Hydro toolset was developed by Dean Djokic, Zichuan Ye, Christine Dartiguenave, Sreeresh Sreedhar, and Amit Sinha at ESRI, and by Tim Whiteaker at the Center for Research in Water Resources. Lori Armstrong provided constant encouragement (and great parties!). Mike Zeiler drew beautiful software diagrams. Claudia Naber, Savitri Brant, Edith Punt, and Tiffany Wilkerson did a great job of producing this book.

It has been a privilege to lead the design and development of Arc Hydro. I hope you will enjoy this book and benefit from our work.

David R. Maidment
Center for Research in Water Resources
University of Texas at Austin

Creating and documenting a generic geographic data model, relevant for widespread use, is a perilous task. It can be likened to a journey of exploration—the designer sets off into a strange world of database design, UML, theoretical geography, and object-oriented jargon to find and describe the "best model." The designer is armed (hopefully) with a deep understanding of the domain, innate common sense, and the drive to get the job done right.

The designer must avoid two great perils: the desert of oversimplicity ("rivers are blue polylines") and the miry swamps of complexity ("let's talk about the meaning of springs—if a fumarole is also a source of hot water, is it a kind of spring or a . . .). Our intrepid designer will get no end of advice from others throughout the journey. For some reason, data model design is a subject that nearly everyone feels they understand, and advice is freely offered. I have found that you don't have to be a rocket scientist to confidently speak about the ontology of rockets or to draw a UML diagram of a Saturn V booster. Most of this advice is useless, but in the metajargon of information modeling, it can sound convincingly precise—"since river objects are associated with their left bank objects and their right banks, island objects must have associated left, right, upstream, and downstream banks. . . ."

The job of developing a data model is not a solo activity. By its very nature, data model design involves other people, both as members of the design team and as potential users of the resulting model. This can lead to a poorly conceived political approach to the design process, with all the worst aspects of the "design by committee" paradigm. The development of a successful, working model is a creative activity, like making a movie or designing an automobile. To be successful, the captain of the design expedition needs to be a responsible and wise leader of people, not simply a process facilitator (at one extreme) or a design dictator (at the other).

In this volume, Dr. David Maidment reports the results of a successful expedition to define and document a GIS data model for hydrology and hydrography (known to most of us as "the blue things on the map"). It is interesting in itself as a useful data model. It is also interesting as a template or example of "good GIS data model design." I have watched the progress of this work since its inception and I believe that there are a couple of reasons for this success.

First, it was always clear that the hydro data model had to do something and not just represent knowledge. The model was designed and constantly evaluated to support the derivation of information products from the data. Specifically, the model needed to directly support both cartographic and water resources applications. The core idea was to "define a simple model that would simultaneously serve as the basic hydro layer on a GIS and also serve water resource applications." The model was continually validated against these practical requirements. Reliance on these functional criteria allowed the designers to avoid the "descriptive geomorphology" swamp (attempting to tease out definitions for intermittent stream, swale, holding pond vs. reservoir, etc.). When considered in terms of the functional requirements for cartography and water resource modeling, these distinctions become mostly irrelevant. In fact, the model simply defines two kinds of network lines—flowlines and shorelines. Other descriptive properties can be associated with these objects, but they are not central to the definition of the model.

The second foundation for the success of the model is the process used to create the model. Dr. Maidment has always been very open to involving a wide variety of interested parties in the design and review of the model through workshops, a consortium of users and developers, and by sharing draft documentation and software. He has been generous in sharing credit and creating an intellectual community around the work, involving many different people and perspectives. However, he clearly has taken responsibility for ensuring that the design has intellectual integrity and that it works.

It has been a great pleasure to be associated with this project. I think that this design will not only serve the hydrologic community of GIS users (a surprisingly diverse and pragmatic lot), but also as a representative example of excellent GIS data model design.

Scott Morehouse
Redlands, California
May 22, 2002

How to read
this diagram

Arc Hydro Data Model

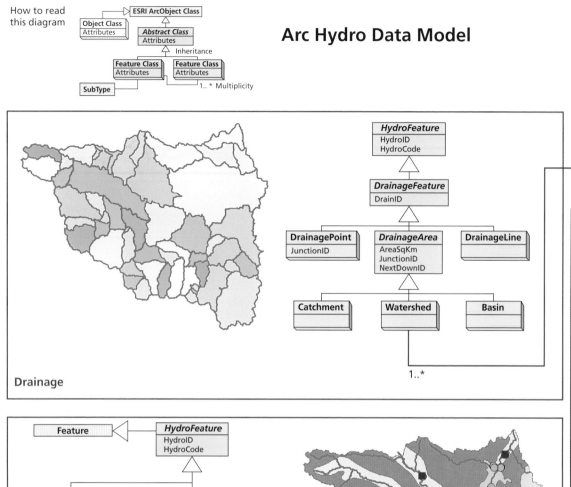

Drainage

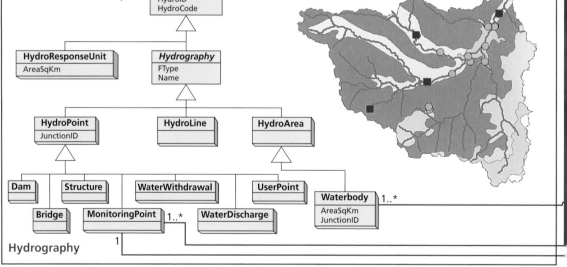

Hydrography

GIS in Water Resources Consortium

ESRI

Network

SimpleJunctionFeature

HydroJunction
- HydroID
- HydroCode
- NextDownID
- LengthDown
- DrainArea
- FType
- Enabled
- AncillaryRole

1
1
1

ComplexEdgeFeature

HydroEdge
- HydroID
- HydroCode
- ReachCode
- Name
- LengthKm
- LengthDown
- FlowDir
- FType
- EdgeType
- Enabled

EdgeType
Flowline
Shoreline

HydroFeature
- HydroID
- HydroCode

SchematicNode
- FeatureID

SchematicLink
- FromNodeID
- ToNodeID

Object

HydroEvent
- ReachCode

HydroPointEvent
- Measure

HydroLineEvent
- FMeasure
- TMeasure
- Offset

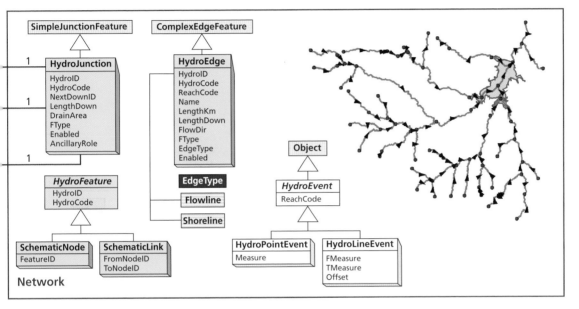

Channel

HydroFeature
- HydroID
- HydroCode

ChannelFeature
- ReachCode
- RiverCode

ProfileLine
- FType
- ProfOrigin

1

CrossSection
- CSCode
- JunctionID
- CSOrigin
- ProfileM

Object

CrossSectionPoint
- CSCode
- CrossM
- Elevation

1..*

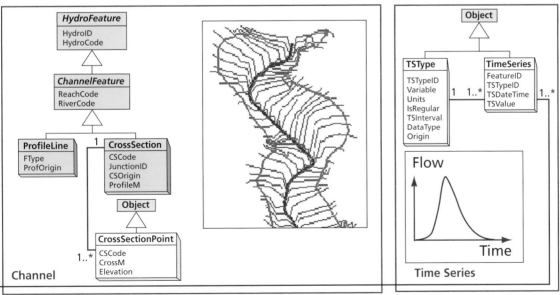

Time Series

Object

TSType
- TSTypeID
- Variable
- Units
- IsRegular
- TSInterval
- DataType
- Origin

TimeSeries
- FeatureID
- TSTypeID
- TSDateTime
- TSValue

1   1..*   1..*

Flow

Time

## chapter one

Why Arc Hydro?

David Maidment, University of Texas at Austin

Water is fundamental to human life and the functioning of the natural environment. The scope and scale of water resources problems makes geographic information system (GIS) software a powerful tool for developing solutions, and the advent of ESRI® ArcGIS™ has created an opportunity to rethink the way that water resources data is represented in GIS. The result is Arc Hydro— an ArcGIS data model for water resources. Arc Hydro opens the way to building hydrologic information systems that synthesize geospatial and temporal water resources data to support hydrologic analysis and modeling.

Water and life

The health of a community's water systems is naturally a concern of its citizens. They expect to be protected from floods, to have a reliable water supply, to swim and fish in their rivers and lakes. But nature can be cruel as well as kind. Floods can devastate communities, destroying in hours the labor of a lifetime. There are always conflicts among people who share water resources, conflicts that become critical during droughts when there is not enough water to meet everyone's needs. Population growth and economic development can degrade water quality and the living environment for wildlife. All these pressures exist in Texas, a state where the recurring cycle of flood and drought is a way of life. The Guadalupe River basin in Texas is used throughout this book as an example application of the Arc Hydro data model.

Rainfall in Texas was lower than normal in the autumn of 1995, the beginning of what would become the 1996 Texas drought. Over the winter and spring, the dearth of rainfall continued, and the rains that normally nourish seedlings and replenish water tables in April and May never arrived. By June, south Texas was already in the midst of drought. In the Guadalupe River basin, experts called the weather an "extended dry spell," but admitted there were signs that things could evolve into a "disaster." Reserves in reservoirs and aquifers continued to fall, crops withered, livestock died, and the 1996 drought ended up costing Texas over $1.5 billion. And this drought wasn't even a particularly extended one—the great drought of the 1950s, which lasted for six years, is etched in the memory of any Texan who lived through it.

As counties across Texas clamored for relief in 1996, then-governor George W. Bush contacted state water officials. How much water do we have, how much are we using, and how much do we need? These intelligent, sensible questions had the potential to solve problems if they could be answered, but state water officials did not have adequate information to provide answers. Recognizing these shortcomings, the Texas Legislature in 1997 launched an ambitious plan to overhaul water planning in the state, and to construct detailed, statewide data layers of geospatial

U.S. Agriculture Secretary Dan Glickman has called drought an even more insidious natural disaster than hurricanes, floods, or tornadoes "because it happens very slowly. Its impact is more long range, far reaching."

information on land-surface terrain, soils, land use, and stream hydrography. Water availability in Texas is now managed using this digital geospatial infrastructure instead of stacks of paper maps. The Arc Hydro data model described in this book is used to structure the geospatial information for Texas' water availability modeling.

The 1996 drought was hardly over when a severe storm in October 1998 caused a huge flood in south Texas. Flood flows in the lower Guadalupe basin were more than twice as large as any recorded since stream-gaging stations were installed on the Guadalupe River during the 1930s. The October 1998 flood killed more than 40 people and caused $2.1 billion in damages, yet numbers and statistics don't tell the human story. When a flood recedes from a home that has been under 23 feet of water, all that can be done is to bring in a bulldozer and clean off the concrete foundation slab; nothing else remains.

The cleanup from the October 1998 flood was still going on when Tropical Storm Allison hit Houston in June 2000. A total of 77,000 buildings were inundated there in what became the most costly urban flood disaster in the history of the United States. The story continues. It is at times like these that understanding and managing water resources becomes so much more than understanding and managing information—it becomes life and death. The question is not whether the next flood will occur, but where and when will it happen?

Accurate floodplain mapping is needed to be able to estimate flood risks precisely. One of the reactions to Tropical Storm Allison was a complete remapping of the floodplains of Houston

"Nature is devastating," said Governor George W. Bush. "It was only three months ago that we were praying for rain, and now in Texas we've got too much rain. It happened so quickly and so suddenly."

using LIDAR (Light Detection and Ranging), a remote sensing technique from aircraft that produces highly detailed and precise maps of land-surface terrain. Chapter 5 in this book shows how Arc Hydro can be used to process terrain information for floodplain mapping. Precise terrain information is also being used to design new flood channels and pipelines to carry away flood waters from the next severe storm in Houston.

In the United States, the Nexrad radar rainfall system provides maps of storm precipitation updated every few minutes. There would be a very significant benefit for public safety if these radar rainfall maps could be quickly translated into anticipated flood inundation maps during storm events, so that citizens could be better warned to stay away from roads that cross swollen rivers and streams, since being swept away in a vehicle causes many deaths in floods in Texas. Being able to create real-time flood inundation maps requires connecting map data on ground conditions with a time sequence of rainfall maps and with flood simulation models to produce a time sequence of flood maps on the ground. Chapter 7 shows how time sequences of Nexrad radar rainfall map data can be incorporated into Arc Hydro.

Concern about water resources is not limited to the impacts of floods and droughts. Citizens want livable cities, where urban growth and development can coexist with a clean environment and healthy ecosystems. Water-quality management in cities requires solutions that assess how regulation of land development will affect water quality and ecological resources. What combination of pollution prevention structures, education, and regulation best serves to enhance the quality of the urban water environment?

From a larger perspective, in the United States the goal of the 1972 Safe Drinking Water Act was to make the nation's waters fishable and swimmable. After thirty years of pollution prevention efforts, mostly aimed at controlling point sources of pollution, many of the nation's waters

Photography used by permission of Francisco Olivera

Urban hydrologic systems pose many challenges not seen in rural systems.

are still not fishable and swimmable because of nonpoint sources of pollution, such as runoff from farmlands.

To rectify this problem, the U.S. Environmental Protection Agency (EPA) has embarked on an ambitious program called Total Maximum Daily Load, whose goals include setting regulatory standards on the quality of water within natural water bodies, rather than regulating just the outflows of wastewater discharge pipes. Trying to regulate the quality of natural water bodies requires a much more comprehensive understanding of the impact of all types of pollution sources on receiving water quality than has been required in the past. The EPA has created a decision support system called Basins, built using ArcView® version 3, to integrate geospatial and temporal water resources information, and water-quality models for improved water-quality management. The Arc Hydro data model presented in this book can be used to extend the Basins system so that it can be implemented in a more robust and effective manner using ESRI's new ArcGIS technology.

These examples of the importance of water issues in Texas and the United States are not unique. The same issues are faced in other countries, and better solutions to information management for water resources are needed throughout the world.

GIS for water resources

Hydrologists use many data sources to assess water quality, determine water supply, prevent flooding, understand environmental issues, and manage water resources. During the 1990s, GIS emerged as a significant support tool for hydrologic modeling. In particular, GIS provided a consistent method for watershed and stream network delineation using digital elevation models (DEMs) of land-surface terrain. Standardized GIS data sets for land cover, soil properties, gaging station locations, and climatic variables were developed, and many of these data sets are published on the Internet. GIS data preprocessors were developed to prepare input data for water flow and water-quality models. GIS is now accepted as a useful tool for assembling water resources information, and the community of water resources and GIS specialists who are familiar with these tools and data sets is growing.

While much progress has been made with the application of GIS in water resources, and the creation of national, regional, and local data sets, many challenges remain:

- It is essential for hydrologists to depict the flow of water through the landscape in space and time. How can time series data on water flows and water quality be integrated with geospatial data describing the water environment? This is perhaps the most critical challenge.
- Water management agencies are building GIS data sets to support their operations, but often each particular project or group develops its own data. How can an agency produce a common geospatial data infrastructure to support a range of water resources projects?
- In water resources studies using several hydrologic models, each model usually operates independently with its own GIS database. How can a data framework be developed to support several linked hydrologic models?

5

- As digital elevation data sets for land-surface terrain become more refined, their cell size decreases and the number of cells increases to the point where studying a large region with a single digital elevation model data set is cumbersome and impractical. How can terrain information be subdivided into subareas for analysis, and the results merged to form a watershed data set for the whole region?

- Triangulated irregular network (TIN) data is used to describe river channels and floodplains, but their degree of detail also becomes overwhelming as the length of river being studied increases. How can a vector description of stream channel shape be built from a TIN surface for each river segment, and then the results combined to form a continuous description of channel shape for the whole length of the river?

- Ecologists and geomorphologists conduct detailed surveys of river reaches to identify substrate and organism types. How can the information gathered in these surveys be stored in a structured way so that it is linked to the geometrical information about the stream channel shape?

- Standardized hydrography data sets are being produced to store the "blue lines" describing streams, rivers, lakes, and coastal water bodies. How can stream hydrography be linked with independently delineated watershed areas defining the land areas draining to these streams?

- A large number of point features are associated with river networks, including gaging stations, hydraulic structures, intakes for water supply systems, discharge points for wastewater treatment plants, bridges, and locations where rivers cross administrative or aquifer boundaries. How can this information be addressed along the river network so that it is easy to tell what is upstream and downstream of a particular location?

- Urban hydrologic systems pose many challenges not seen in rural systems. In particular, the drainage areas of storm sewer systems may be quite different than the drainage areas of surface streams in the same region. How can this discrepancy be accounted for?

The common theme among these challenges is the need for integration: integration of different types of GIS and water resources data, integration of data and modeling, integration across spatial scales. The need is clear, but not so the means for accomplishing it. Some requirements for an integrated water resources data system are:

- All data must be held in a common geospatial coordinate system.

- The primary structure used for representation of a large region must be vector data (points, lines, and areas), supported by raster and TIN surface data where necessary.

- Relationships among geographic features in different data layers are needed to trace water movement from feature to feature through the landscape.

- It is critical to be able to link geospatial information describing the water environment with time series information about water measurements to form a complete information system for water resources.

ArcGIS and Arc Hydro

The GIS software packages used most widely throughout the world are the ArcInfo™ and ArcView systems developed by ESRI in Redlands, California. ArcInfo was originally developed in 1980 to use a combination of vector data (points, lines, and areas) with tabular attributes, and was later extended to include surface modeling using square-cell raster grids and triangulated irregular networks. ArcView was developed in the early 1990s, initially as a simple viewing software for GIS data, then later expanded to support spatial analysis and modeling. Special programming languages, ARC Macro Language (AML™) for ArcInfo and Avenue™ for ArcView, were used to permit customization of the GIS for particular applications, such as to create GIS preprocessor modules for water resources simulation models.

ESRI recently reengineered its entire GIS software system to construct a new GIS more closely following current information technology and software engineering standards. This new product, ArcGIS, is presently being distributed to the ESRI user community. ArcGIS comes in several variants depending on the degree of functionality required by the user, with ArcView remaining the entry-level version of ArcGIS for data viewing, querying, and analysis, and ArcInfo as the high-end version of ArcGIS for data creation and sophisticated operations. Microsoft® Visual Basic® is the standard interface language for ArcGIS, just as Microsoft uses Visual Basic as the interface language for Microsoft Excel, Microsoft Access, and other application programs.

As part of the ArcGIS development effort, ESRI initiated efforts to show how ArcGIS can be customized for particular applications of GIS using specially designed data models. The water resources data-modeling effort was undertaken in association with the Center for Research in Water Resources (CRWR) of the University of Texas at Austin, and CRWR and ESRI together formed a GIS in Water Resources Consortium to involve representatives from industry, government, and academia in the water resources data model development. The water resources data model was discussed at several national conferences held in the United States and in many other communications among members of the Consortium community. A series of prototype data models was built and tested at CRWR until the current form was reached. The result is formally called the ArcGIS Hydro data model, but is informally known as Arc Hydro, as it is in this book.

Arc Hydro is a geospatial and temporal data model for water resources that operates within ArcGIS. Arc Hydro has an associated set of tools, built jointly by ESRI and CRWR, that populate the attributes of the features in the data framework, interconnect features in different data layers, and support hydrologic analysis. For more information, see the Web site for the GIS in Water Resources Consortium at www.crwr.utexas.edu/giswr, or ESRI's Web site for ArcGIS data models at arconline.esri.com/arconline/datamodels.cfm.

Arc Hydro is a data structure that supports hydrologic simulation models, but it is not itself a simulation model. Hydrologic simulation is accomplished by exchanging data between Arc Hydro and an independent hydrologic model, by constructing a simulation model attached to Arc Hydro using a dynamic linked library, or by customizing the behavior of Arc Hydro objects. This is described further in chapter 8.

The experience of the authors of this book has shown that it is best to begin with the simple version of Arc Hydro called the Arc Hydro framework, then add additional classes and attributes to the framework model as the need dictates. Arc Hydro is based on concepts and ideas that can be adapted and developed to suit individual applications. Arc Hydro supports basic information management tasks for water resources, and serves as a point of departure for further elaboration to fit particular circumstances.

Arc Hydro describes natural water systems, not constructed water pipe systems such as water supply, wastewater collection, and storm sewer systems. A separate data model called ArcGIS Water Utilities has been developed by ESRI to describe water pipe infrastructure systems. It should also be noted that, while Arc Hydro data structures lend themselves to the depiction of data from aquatic ecology and geomorphology, those subjects were not a prime focus during the Arc Hydro design effort. Moreover, Arc Hydro has no explicit data structures for aquifers or other groundwater features; it is focused on the description of surface water hydrology and hydrography.

While the need for describing data in a formalized manner may be obvious to a GIS specialist, a hydrologist unfamiliar with this terminology may ask: why do I need all of this structure and formality? And for a project on a small area where most of the data is gathered by hand or from paper maps, there may be little need for a data model. But when projects cover larger areas, make use of several different sources of GIS data, and involve running complicated hydrologic models, a greater degree of formality in the way the information is structured provides more systematic and efficient project execution, and the potential to reuse the data on subsequent projects in the same area. In the future, hydrologists will increasingly rely on GIS data, and Arc Hydro provides a standardized way of describing that data so that it can be used consistently and efficiently.

Hydrologic information systems

From a traditional water resources point of view, the term "water resources data" has primarily meant time series data on observations of water resources phenomena, including rainfall, streamflow, water quality, and climate. As the use of GIS in water resources has increased, the concept of water resources data has broadened to include geospatial data describing the water resource features of the landscape. Although time series data on water properties and geospatial data on the water environment have been thought of as quite different things, they are really both just information sources that the hydrologist wants to use.

The design of Arc Hydro revealed that it is now possible to define a "hydrologic information system," which is a synthesis of geospatial and temporal data supporting hydrologic analysis, modeling, and decision making. This is a very exciting new concept because rather than simply applying GIS in water resources, what is being created is the foundation for a new way of thinking about how information technology can be used to support water resources.

There is a significant synergy between geospatial and temporal water resources information that up to this time has been difficult to capture because the geospatial and temporal data have been held in different formats and archiving environments. Arc Hydro provides a simple but systematic way to link time series data on water measurements to geospatial data on the locations where measurements are made. This happens within a single information system so that

Hydrologic information system

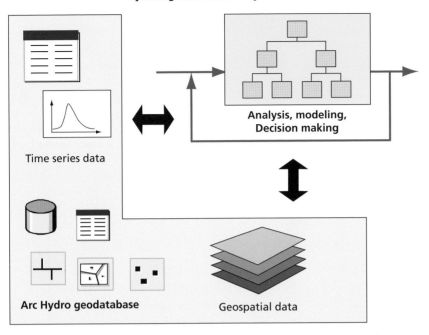

A hydrologic information system connects time series and geospatial data
with hydrologic analysis and modeling.

the movement of water throughout the stream and river network can readily be traced. Moreover, since any type of time series data can be linked to any geospatial feature in Arc Hydro, it is possible within ArcGIS to construct time-sequenced maps of rainfall and the water conditions of the landscape.

As an example of the combination of geospatial and temporal water resources data, a prototype application called Arc Hydro USA was built to describe the river and stream network, water bodies, and drainage areas of the entire continental United States. In addition, Arc Hydro USA contains the locations of some 18,000 U.S. Geological Survey streamflow gaging stations and the daily streamflow time series record for those stations in the Hydro Climatic Data Network, a set of long streamflow records selected to reflect the streamflow history of the nation. All this is contained in a single Arc Hydro geodatabase stored in Microsoft Access and readily usable in ArcGIS on a laptop computer. And because Microsoft Access can be directly linked to Microsoft Excel, Arc Hydro data can be directly viewed and manipulated within Microsoft Excel spreadsheets without going through ArcGIS at all. These are remarkable accomplishments and promise the opening of a new era of information management in water resources.

A state, a regional water management organization, or a city can also apply Arc Hydro to assemble a comprehensive water resources management database for its region of jurisdiction. This application involves most or all of the feature classes in the Arc Hydro data model, but the

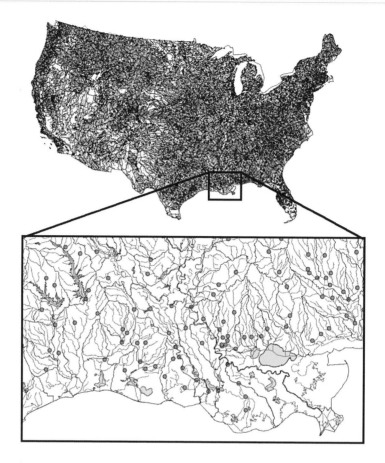

Arc Hydro USA, a prototype water resources geodatabase for the continental United States. Over 7,000 USGS streamflow gaging stations plus daily streamflow time series records are stored in a single Arc Hydro geodatabase.

geographic extent is smaller than in the national example just described. In this book, several examples are presented of how Arc Hydro can be applied for the Guadalupe River basin in Texas, illustrating how a regional data model can be built.

A smaller organization or an individual can take an existing Arc Hydro database and add more information to it, such as more detailed drainage areas, and stream channel definitions for particular water resources studies. A city or regional water management agency can build and maintain a core geodatabase, and consultants can add detail when doing studies in particular areas. Arc Hydro can even be applied at the scale of a small urban subdivision to study runoff patterns through building lots and storm sewers. Arc Hydro is carefully designed to be a simple data structure that can readily be applied to study water resources problems at any spatial scale.

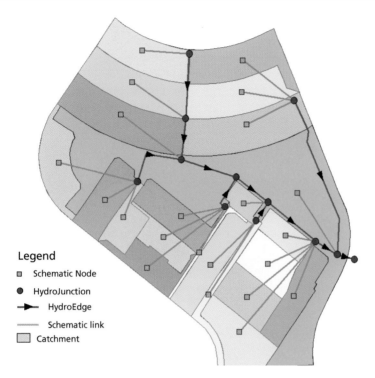

Arc Hydro applied to an urban subdivision. The HydroEdges are storm sewers, the HydroJunctions are storm sewer inlets, the catchments are individual building lots and a grassy swale draining through the middle of the subdivision (the bright green area). The linkage between the catchments and the storm sewer inlets is symbolized with SchematicLinks and SchematicNodes.

Outline of this book

To describe and implement the Arc Hydro data model, we use:
- An analysis diagram summarizing graphically the taxonomic grouping and hierarchical arrangement of classes and objects in the data model. This is included in the preface of this book.
- A data dictionary for the database tables where the data is stored. These are summarized at the end of chapters 3 to 7.
- A vocabulary attached as a glossary appendix to this book that defines the terms used in the analysis diagram.
- A set of Arc Hydro tools to populate the attributes of the features in the data model, inter-connect features in different data layers, and allow operations to be performed on them. The current Arc Hydro toolset and instructions are available on the Arc Hydro CD–ROM at the back of this book, which also includes an example Arc Hydro application involving the Guadalupe River basin in Texas.

Chapter 1 of this book has served to introduce the use of GIS in water resources and some motivations for the development of Arc Hydro. Chapter 2 presents the Arc Hydro framework, a simplified version of Arc Hydro designed to support geospatial information on stream and river networks, watersheds, water bodies, and monitoring points where water is measured. The method by which the Arc Hydro framework is built from a generic set of spatial objects called ArcObjects™ is also described. Chapters 3 through 6 show how to build the hydro network, delineate drainage areas, describe channel features, and assemble map hydrography data sets to form a comprehensive description of geospatial data for hydrology. Chapter 7 focuses on the theme of time series, and shows how Arc Hydro represents water resources time series as a set of data values, each value indexed by the spatial feature to which the data value applies, the type of time series data, and the date and time of the value. Chapter 8 outlines how geospatial and temporal Arc Hydro data can be used together to support hydrologic analysis and modeling using various interfaces, such as Visual Basic programs, Microsoft Excel, Microsoft Access, and ArcGIS. Finally, chapter 9 describes the technical steps involved in implementing Arc Hydro within ArcGIS.

Arc Hydro framework

David Maidment, University of Texas at Austin
Scott Morehouse, ESRI
Steve Grise, ESRI

The Arc Hydro framework is a simplified version of Arc Hydro. The framework stores information about the river network, watersheds, water bodies, and monitoring points. Additional components may be assembled around the framework to form a more complete Arc Hydro data model. This chapter describes the design of Arc Hydro, the ArcObject structure of ArcGIS, and the key concepts used in the Arc Hydro framework.

Water resources environment

Arc Hydro is designed to represent data from hydrography and hydrology, thereby creating a basis for obtaining a deeper understanding of surface water systems.

Hydrography

All topographic maps contain hydrography, the "blue lines" that signal to the map reader the meandering path of a river, the shoreline of a lake, or the location of the coastline. Even very old maps depict rivers and coastlines.

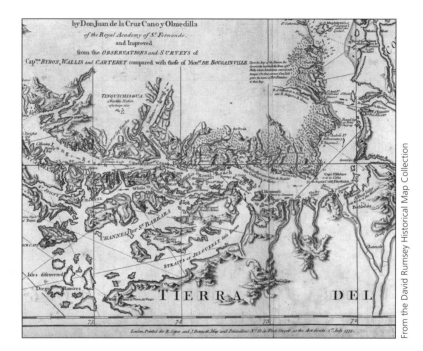

From the David Rumsey Historical Map Collection

A map of the straits of Magellan created in 1776 by Thomas Jefferys.

Webster's dictionary defines hydrography as "the description, study and charting of bodies of water, such as rivers, lakes and seas." Topographic maps also contain hypsography, the contour lines showing the elevation and shape of the land surface. Because the drainage of water through the landscape is a powerful force in shaping the land-surface terrain, hydrography and hypsography are closely related. The character of the land use, whether urban, agricultural, or rural, also affects water flow through the landscape. Map information is critical to determining links between the land and water systems.

In recent years, an important transformation has taken place as GIS has been used to convert paper map products into digital data layers. The Internet is serving as a convenient mechanism

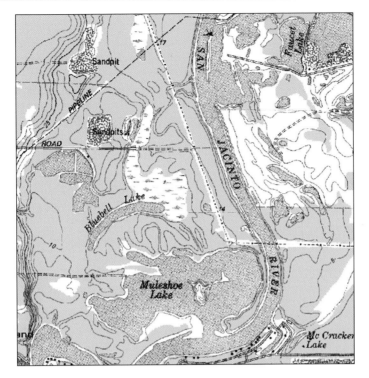

Digital raster graphic image of map hydrography and hypsography.

Hydrology
The natural partner of the hydrographic description of water bodies is the study of water movement through them. This is the domain of hydrology, which Webster's defines as "a science dealing with the properties, distribution, and circulation of water on the surface of the land, in the soil and underlying rocks, and in the atmosphere." Hydrologic data is obtained from field monitoring stations, such as streamflow and rain gage recorders and climate stations, that record through time the variations in water flow and related quantities. The physical laws that govern water flow and the transport of constituents in water are captured in dynamic computer models, calibrated by comparing them with observational data. These models attempt to simulate how water moves through the landscape and how water properties change as it does so. Obviously, it is necessary to characterize the nature of the landscape with parameters and data to make such simulation possible.

for widespread distribution of such information. Recent years have also witnessed a dramatic increase in the capacity to characterize the landscape by using remote sensing. More satellite and aerial photography, and products derived from this imagery, are available in useful forms. The degree of precision and detail of the digital description of the natural landscape is continually expanding.

Hydrology differs from hydrography in the scope of its focus. Hydrology deals with atmospheric, surface, and subsurface water, while hydrography is mostly concerned with surface water. Also, hydrology is primarily concerned with defining the properties and movement of water, while hydrography defines the nature of the environment through which water moves. Hydrologic data is measured at a few points in space but varies rapidly through time. GIS description of hydrography and related data is spatially intensive and changes little in time. The common ground where hydrography and hydrology meet is in the description of surface water systems.

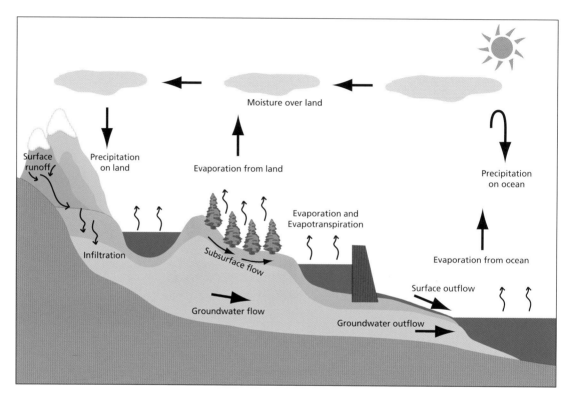

Hydrologic cycle—the circulation of the waters of the earth.

Building a geographic data model

A data model is simply a method for describing a system using a structured set of data. For example, the set of variables used in a computer program is the data model for the program. Data models provide an orderly way to classify things and their relationships. A geographic data model is a representation of the real world that can be used in a GIS to produce maps, perform interactive queries, and execute analyses. Geographic data models define a vocabulary for

Thematic layers of the Arc Hydro data model

Layer Streams
Map use Cartography and stream analysis
Data source Usually mapped by government mapping and resource agencies
Representation Edges and nodes for the stream network, polygons for lakes
Spatial relationships Each edge has a flow direction and flows into another edge or sink
Map scale and accuracy A typical map scale is 1: 24,000, locational accuracy is about 10 meters
Symbology and annotation Streams are drawn with blue lines with varying weights and patterns with line color, weight, and style

Layer Hydrographic points
Map use Gage stations on a stream network and features such as dams
Data source Usually mapped by government mapping and resource agencies
Representation Junctions, network flags, and points on a stream network
Spatial relationships Points can be related to junctions on the network
Map scale and accuracy A typical map scale is 1: 24,000, locational accuracy is about 10 meters
Symbology and annotation Typically drawn with colored circle markers by type

Layer Drainage areas
Map use Drainage areas are used to estimate water flow into rivers
Data source Derived from digital elevation models
Representation Polygon with points at drainage outlets
Spatial relationships Each drainage area covers a stream section
Map scale and accuracy A typical map scale is 1: 24,000, locational accuracy is about 10 meters
Symbology and annotation Shaded polygons can depict catchments or watersheds

Layer Hydrography
Map use The hydrographic layer in topographic maps
Data source Mapped by government mapping agency
Representation Point, line, polygon, and annotation for water features
Spatial relationships Streams feed rivers, rivers flow into lakes or oceans
Map scale and accuracy A typical map scale is 1:24,000, locational accuracy is about 10 meters
Symbology and annotation National cartographic standards are applied to water features

Layer Channels
Map use Hydraulic analysis
Data source Derived from surface model or land surveying
Representation Cross sections and longitudinal profiles along a river channel
Spatial relationships Cross sections are perpendicular to flowlines
Map scale and accuracy A typical map scale is 1:2,400 with location accuracy about 1 meter
Symbology and annotation Channels, flowlines, and cross sections shown with graphs

Layer Surface terrain
Map use Deriving streams and drainage areas, also cartographic background
Data source Digital elevation models
Representation TIN surface model or raster with elevations
Spatial relationships If raster, each cell has an elevation; if TIN, each face joins to form surface
Map scale and accuracy A typical map scale is 1:2,400 with location accuracy about 1 meter
Symbology and annotation Elevation is usually shown with graduated colors

Layer Rainfall response
Map use Overlayed with rainfall grid to estimate flood or drought conditions
Data source Derived from combining layers such as soil, vegetation, and land use
Representation Polygon
Spatial relationships Polygons tesselate an area
Map scale and accuracy A typical map scale is 1:24,000, locational accuracy is about 10 meters
Symbology and annotation Polygons can be shaded in proportion to rainfall response values

Layer Digital orthophotography
Map use Map background
Data source Aerial photogrammetry and satellite collection
Representation Raster
Spatial relationships Pixels tesselate the area imaged
Map scale and accuracy Pixel resolution typically is 1 to 2.5 meters or better
Symbology and annotation Tone, contrast, and balance of grayscale or color presentation

describing and reasoning about things located on the earth. By agreeing on a common vocabulary, practitioners can readily share data, tools, and analyses. Geographic data models also provide ways to model systems of behavior describing how things in the landscape interact with each other.

There are two basic approaches to building a geographic data model: inventory and behavioral. These two approaches can yield two very different data models, and in practice, a data modeler blends both approaches—a task that is easier said than done!

Inventory approach to data modeling

In an inventory approach we consider all those things of interest in the landscape and formally define them by describing their location, properties, and individual behaviors. This involves grouping like things into classes or data layers, and it results in the classic GIS database picture: a stack of data layers describing different kinds of spatial features in a given geographic region.

In water resources, there are two sources of inventory data: map hydrography and tabular data inventories. Map hydrography is depicted by the "blue lines" on a topographic map: the points, lines, and areas that describe water features. Independently of maps, many organizations have prepared tabular data inventories of items like gages, dams, and bridges, where the location of the feature is recorded simply by latitude and longitude coordinates. These coordinates can be used to create layers of point features so that a complete water resources data inventory can be compiled in the map environment.

An inventory approach is an excellent way to describe the water features and their location in the landscape.

Behavioral approach to data modeling

In a behavioral approach we define a system of behavior, look for pertinent features to describe the behavior, then define how the behavior of individual features interacts with others to define the behavior of the whole system. This is the classic water resources modeling approach. Its emphasis is not on describing the environment in exhaustive geographic detail, but rather in creating a schematic view of the landscape that highlights the features important to a particular water resources model or analysis.

Water resources are modeled for many purposes, such as to study flooding, water quality, or water supply, or simply to understand the hydrology of a given region. Each of these purposes involves modeling different aspects of the behavior of water, as described by physical laws and sets of equations. Solving the governing equations of a water resources model can be a complex task in its own right, and a geographic data model cannot describe all the different behavioral concepts that such models and analyses require—that is a task for water resources specialists. Yet every water resources simulation model also contains a simplified view of the water features of the landscape, which can be thought of as its elementary geographic data model, expressed in the arrangement and data values contained in the input file for the water resources model's operation.

Beneath all the modeling there is a fundamental fact: a river is a river is a river, regardless of the many ways in which rivers are depicted in water resources models. Likewise, land-surface topography defines the broad patterns of drainage in the landscape, from the land surface to streams, and rivers, and then to the ocean. These drainage patterns exist independently of the way they are portrayed in individual water resources models.

Suppose that we adopt as a behavioral model the simple concept: "follow a drop of water from where it falls on the land, to the stream, and all the way to the ocean" (Hirsch 2001). Of course, true water resources modeling is a good deal more complicated, but the virtue of the raindrop model is that it allows an orderly procedure to be established for describing how water flows from one feature to another through the environment.

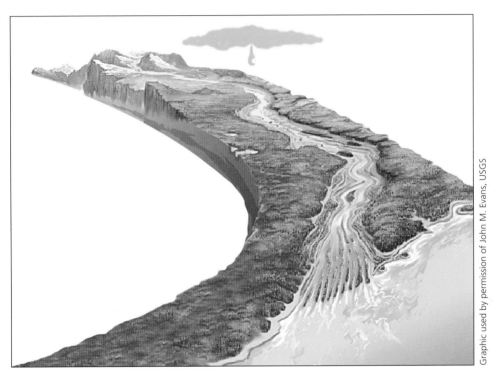

Graphic used by permission of John M. Evans, USGS

A behavioral model representing the path of a raindrop through the landscape

Database integration

The task of tracing the path of a raindrop as it flows through the landscape involves knowing the drainage system on the land surface, the natural flow system in streams, rivers, lakes, and constructed flow systems such as ditches, pipes, and canals. The effects of dams, bridges, and pumping systems on the flow patterns also have to be accounted for. This requires an additional kind of functionality, namely knowledge of the connectivity among features in different data layers. ArcGIS supports a formal network connectivity model of points and lines in its network analysis system, and Arc Hydro adds relationships between network junctions and related hydrologic features, such as watersheds, water bodies, and monitoring points. This "feature-to-feature" connectivity is an example of an important aspect of ArcGIS: ArcGIS supports "feature-oriented operations" as well as "layer-oriented operations." In other words, besides collecting all the watersheds in one data layer, and all the stream segments in another layer, relationships in ArcGIS can also be established between an individual watershed that drains to a particular stream segment, just as it does in nature. Moreover, these relationships are stored permanently as part of the data structure.

ArcGIS differs in significant ways from earlier GIS systems, including the earlier versions of ArcInfo and ArcView, and these differences are beneficial for Arc Hydro.

ArcGIS features

Three aspects of ArcGIS emerged as critical factors in designing Arc Hydro: networks, the geodatabase, and object modeling.

Network features

The river network of streams, rivers, and water bodies is an easily recognized set of water features on a topographic map. ArcGIS has a special class of more sophisticated point and line features called network features, consisting of lines or edges joined by points or junctions. A network has three components: its geometric data specifies where each point or line is located in

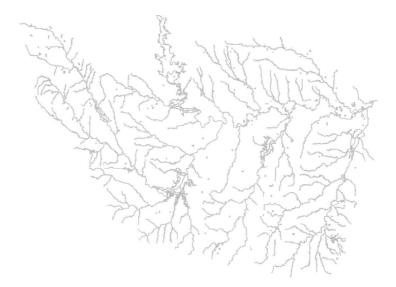

A GIS river network.
The teal-colored speckles represent water bodies. In this case, the rural area is dotted with farm ponds.

x,y,z space, its topological data specifies how the edges and junctions are joined to one another, and its addressing data specifies a location along a river, somewhat like a street address.

Taken together, these properties of network features provide a more robust structure for network representation in GIS than was previously available. This fact will have an important impact on the application of GIS in water resources in the forthcoming decade, comparable in magnitude to the impact of digital elevation models during the 1990s, because it is leading to the construction of extensive and detailed models of stream and river networks, and also piped water flow networks. A GIS river network with centerlines through water bodies permits continuous graphical tracing of water movement through the flow system.

Geodatabases

Data can be stored in digital form in three ways: as flat files, as flat files with a data dictionary, and as a relational database. A flat file is simply a listing or table of the data in text form, for example, a table for daily streamflow discharges over a year, with columns for each month and rows for each day of the month, and summary statistics of monthly flows at the bottom of each column. A flat file with a data dictionary is a more structured data format in which each column or field has a special data type and a header title, each row or record contains a data value, and the result is stored as a binary data file that includes both the data and a dictionary specifying the field data types and titles. The INFO™ files used to store descriptive attributes in ArcInfo coverages and the dBASE® files used for the same purpose in ArcView shapefiles are both flat files with a data dictionary.

Most modern data storage systems are now built on relational databases, such as Microsoft Access and Oracle®. Even water resources data archives, such as the U.S. Geological Survey's National Water Information System, held for decades in specially constructed binary files, are being converted to relational database format. A relational database differs from a flat file with a data dictionary in that all data is stored in a set of tables linked by relationships, which are associations between records in connected tables through values in key fields that the tables share. The user always accesses a view or copy of a table or a set of tables in a relational database, rather than having access to the original data table itself. A relational database is a more rigid and interconnected structure than a set of flat files with data dictionaries.

In the pre-ArcGIS version of ArcInfo, the geospatial coordinate data describing the geographic features (Arc) was held separately from the attribute data describing those features (Info). In ArcGIS, all this data can be loaded into a relational database, so that the geospatial coordinate data of a GIS data layer is stored in just one field in a relational data table (Clark et al. 2001). This special form of a relational database is called a geodatabase. Since the relational database supports permanent relationships between its tables, feature-to-feature connections can be set up among the data layers.

This is critical in meeting the need in water resources for tracing water movement through the landscape. What we want in water resources is the capacity to say "water flows from this catchment to this stream, and from this stream to this lake, and by this stream there is a gage at this location recording the flow, and the current value of the flow is . . ." In other words, we need the capacity to be able to link individual features within GIS data layers such as catchments, streams, lakes, and gages, so that we understand how these features interact with one another. The concept of permanent relationships within the geodatabase allows these connections to be made and retained. Since time series are also being stored in relational databases, the connection between geospatial and temporal data can also be created in the ArcGIS environment.

Object modeling

During the 1990s, object-oriented programming became a standard in the software engineering industry, typified by the Microsoft Office suite of products (Microsoft Word, Excel, Access, PowerPoint®). Objects from one application can be copied and used in another, for example a Microsoft Excel table can be inserted in a Word document. This is possible because the Microsoft Office objects are built using a protocol called the Component Object Model (COM), which specifies how objects should interact with one another, thus enabling a Microsoft Excel table object to behave appropriately when inserted in an application environment different than the one in which it was created.

Visual Basic is the programming language used to build the interfaces to the Microsoft Office products, and a variant of Visual Basic, called Visual Basic for Applications, is used to construct macros and application programs that customize Microsoft Excel and Access to solve particular problems beyond their native capabilities. A remarkable feature of this environment is that it allows application programs in other languages, such as FORTRAN or C, to be attached as dynamic linked libraries. Because of this, water resources simulation models in languages such as FORTRAN can be tightly linked into the system. The ArcGIS software also uses Visual Basic as its interface language, and its objects conform to the COM protocol. This means that ArcGIS objects can be copied into Microsoft Office applications and vice versa, so that an ArcGIS map can be inserted in a Word document or a Microsoft Excel graph. Moreover, data developed in ArcGIS and stored in a Microsoft Access personal geodatabase can easily be viewed in Microsoft Excel, and thus incorporated into hydrologic analysis. It is even possible to work directly with Arc Hydro data without using ArcGIS at all!

Traditional computer languages such as FORTRAN are called procedural programming languages because their programs start at the start and end at the end without interaction with the user. An object-oriented programming language such as Visual Basic is "event driven," which means that the chain of logic followed by the program can be initiated by user-generated events, such as clicking on the screen. Objects interact with one another in complex ways, so software design for object-oriented programming is more complicated than it is for procedural programming. The most critical aspect of this design is the design of the objects themselves, which is what lies at the core of the Arc Hydro development.

ArcObjects and the geodatabase

Arc Hydro is a connected set of objects and features built on top of a generic set of objects and features called ArcObjects that is delivered to the user as part of ArcGIS.

To understand this process, it is necessary to define basic terms. An ArcGIS database is called a geodatabase because it stores data describing geospatial objects in a relational database format. The term "class" denotes a group of objects with similarly defined attributes and behaviors. An object is one member of an object class. Attributes of an object are its descriptive properties,

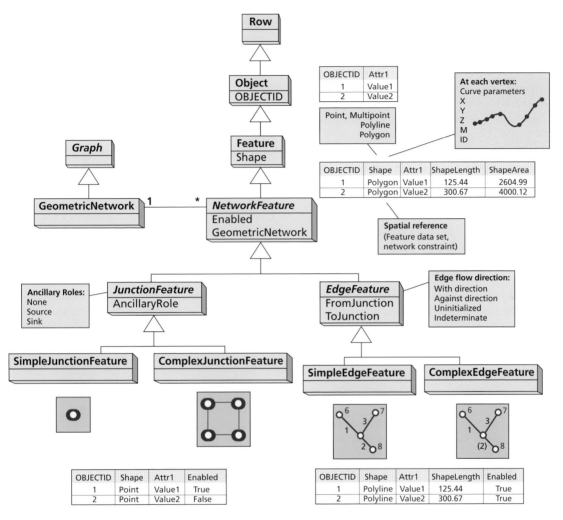

ArcObjects are defined in a Unified Modeling Language (UML) hierarchy.

stored as numbers or text fields in columns of an attribute table, which has one row for each object. The behavior of an object describes how it interacts with other objects, and is defined by program code attached to the object. The most important kind of behavior is the object's interface, which contains the protocol for passing information to and from the object.

23

ArcObjects

ArcObjects are a set of generic ArcGIS object classes used for building geodatabase models such as Arc Hydro. There are three kinds of ArcObjects:

- Objects—These are data tables that store only attributes, such as a time series data table.
- Features—These are data tables that store both spatial coordinates and attributes (points, lines, and areas), such as a stream gage location.
- Network features—These are special points and lines called junctions and edges whose data tables store the connectivity between the junctions and edges, in addition to their attributes and spatial coordinates. Stream segments are represented by network features.

Network features are subdivided into two kinds:

- Simple network features—Formed by a single spatial feature and a single set of attributes.
- Complex network features—Formed by several connected spatial features and a single set of attributes.

As illustrated in the previous UML diagram, ArcObjects are arranged in a hierarchy. All objects lower in the hierarchy inherit the attributes and behaviors associated with the objects above them. This hierarchy is defined using the Unified Modeling Language (UML), a standard language for describing objects developed in the software industry and adopted by ESRI as the basis for the design of ArcGIS. The boxes shown in the diagram have two parts: an upper area with the name of the object class, and a lower area with the attributes of the class, if any. For example, the Feature class has the attribute Shape, which defines whether it is a point, multipoint (several points considered as one feature), polyline (one or more lines considered as one feature), or polygon (a closed set of one or more polylines).

The Feature class is subordinate to the Object class, so each feature inherits the attribute ObjectID, a unique identifier of an object within its class. Besides ObjectID and Shape, more attributes can be added to further describe the features. The NetworkFeature class inherits from the Feature class, so to the standard feature attributes is added the attribute Enabled, which defines whether flow can occur through a network feature. EdgeFeature and JunctionFeature inherit from the NetworkFeature and include additional attributes, such as AncillaryRole, which specifies for Junctions whether the junction is a source of flow to the network, a sink for taking flow from the network, or a regular flow junction that simply transmits flow between edges.

In the Arc Hydro design process, a specialized geometric network was developed that adds additional attributes to the ArcGIS edge and junction features to support water resources operations. This is called the hydro network. During the design process, there was a considerable debate as to whether simple or complex network features should be used to build the hydro network. The decision was made to use simple rather than complex junctions because HydroJunctions don't have complex switching patterns like electrical switch boxes do. However, complex edges were used so that HydroJunctions could be placed on the interior of Hydro-Edges without breaking the edges. These interior junctions can serve as drainage area outlets, reference locations for gages, and points of water withdrawal or discharge.

Arc Hydro features

The first step in customizing ArcObjects for hydrology is to define what makes Arc Hydro features different from any other kind of geographic feature in ArcGIS. In ArcObjects, the ObjectID is a unique identifier of any object within a feature class whose value is assigned at the time the object is created and is maintained permanently thereafter. In Arc Hydro, all features are hydro features and all hydro features carry the following attributes:

- HydroID—An integer attribute that uniquely identifies the feature in the geodatabase
- HydroCode—A text attribute that is a permanent public identifier of the feature

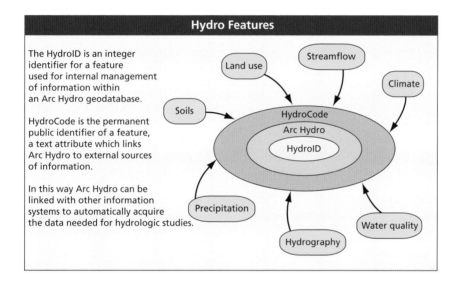

For example, an Arc Hydro application described in chapter 7 reads streamflow time series data for the U.S. Geological Survey's stream-gaging site at the Guadalupe River at Victoria, Texas. Within the resulting Arc Hydro geodatabase, the HydroID of this gage is the integer 12000033, while for the USGS, this gage has a site number of 08176500 (which uniquely identifies the gage among all such stream gages in the United States). Hence the HydroCode of this gage is 08176500, and by knowing this, Arc Hydro can automatically extract streamflow data for this gage through the Internet from the USGS National Water Information System. The HydroID is the label that uniquely identifies features within Arc Hydro, while the HydroCode identifies features with the labels used by other information systems external to Arc Hydro. In this way, Arc Hydro can be linked with other information systems to automatically acquire data needed for hydrologic studies.

Because the HydroID is such an important attribute within Arc Hydro, it is managed carefully. The normal format of a HydroID is a class number followed by a feature number. For example, a HydroID of 12000033 is feature number 33 in feature class 12. The HydroID is defined using a pair of tables, called the LayerKeyTable and the HydroIDTable, that are generated automatically by the Arc Hydro Assign HydroID tool. Each time a new ID is assigned to a feature in a particular class, a counter is updated so that the same HydroID is never assigned again within the given geodatabase. The process of assigning HydroIDs is documented in detail on the Arc Hydro CD–ROM at the back of this book.

The HydroID can be regionalized by storing a two-digit region number before the class number, and in this way, Arc Hydro geodatabases created individually for up to 100 subareas of a region can be merged. All the relationships built uniquely within an individual geodatabase are also valid for the merged geodatabase. Regionalization of data development in Arc Hydro is a powerful tool that enables hydrologists to focus in detail on some areas of a study region and less on other areas, and yet be able to link the results into a single analysis. Regionalization is particularly useful when working with large digital elevation model data sets, as discussed in more detail in chapter 4.

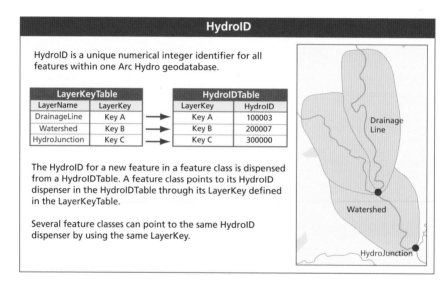

Assigning a HydroID that uniquely identifies a feature in an Arc Hydro database

All hydro features can be associated with any other hydro feature by storing the HydroID of the first feature as an attribute of the second. By this process, drainage areas may be associated with the junctions on the network to which these areas drain, thus defining the correct path of raindrop movement between the land surface and the discharge point on the water flow network. Similarly, time series data can be associated with a particular hydro feature simply by storing the HydroID of that feature with every associated time series data record. The concept that all features in the database are uniquely labeled hydro features is a powerful idea for supporting behavioral modeling, because it means that the database can be considered as an integrated whole rather than as a set of separate data layers.

Within Arc Hydro, all fields used to support HydroID-based relationships are integer fields. This is so because integers are easier to manage than text strings and because database queries operate most effectively using integer variables to index the data values. All Arc Hydro attributes ending in ID (e.g., HydroID, DrainID, FeatureID) indicate an integer identifier, and attributes ending in Code (e.g., HydroCode, ReachCode, RiverCode) indicate a string or text identifier.

Arc Hydro framework

The Arc Hydro framework provides a simple, compact data structure for storing the most important geospatial data describing a water resources system. This framework can support basic water resources studies and models, and can serve as a point of departure for the more extensive data models, that include time series and other Arc Hydro components, as discussed in later chapters of this book. The framework contains information organized in several levels, as depicted in the analysis diagram on the following page:

- Geodatabase—If a personal geodatabase is being used, this is a Microsoft Access .mdb file, or if an enterprise database like Oracle or SQL Server™ is being used, this is a relational database file on a server.
- Feature data set—This is a folder that stores feature classes within the geodatabase. The feature data set has a defined map projection, coordinate system, and spatial extent.
- Geometric network—This is where information that topologically connects hydro edges and hydro junctions is stored.
- Feature class—This is where information on individual geographic features is stored, such as watershed or stream segment information.

- Relationship—This is where features from one class are related to those in another. In this case, each watershed, water body, and monitoring point is related to a hydro junction on the network.

Feature classes of the Arc Hydro framework
The five feature classes shown in blue in the analysis diagram that follows describe the basic geospatial data of a water resources system:
- HydroEdge—A network of "blue lines" describing map streams and water body centerlines.
- HydroJunction—A set of junctions located at the ends of flow segments and at other strategic locations on the flow network. HydroEdges and HydroJunctions are topologically connected in an ArcGIS geometric network, called the hydro network.

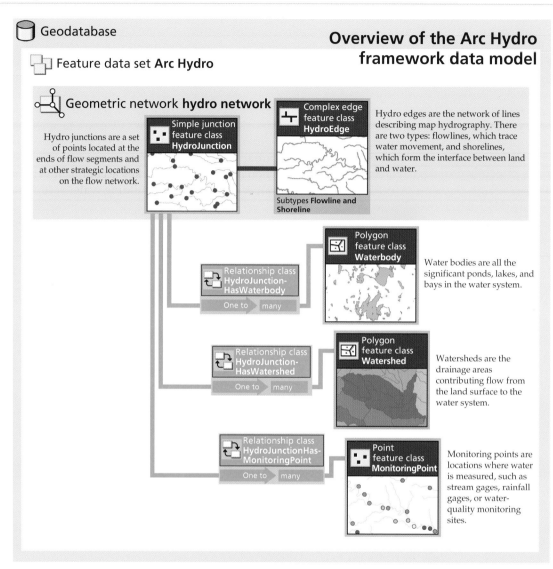

Analysis diagram for the Arc Hydro framework

- Waterbody—The significant ponds, lakes, and bays in the water system.
- Watershed—The drainage areas contributing flow from the land surface to the water system.
- MonitoringPoint—A set of points representing gage locations where water is measured.

The heavy green lines connecting the classes in the diagram depict relationships, which allow associations between individual features and objects in the data structure so that the

MonitoringPoints, Waterbodies, and Watersheds can be linked to the hydro junctions. Each relationship has a multiplicity, and all the relationships shown are one-to-many. One-to-many multiplicity means that one HydroJunction may be associated with one or more features in the related class. For example, two watersheds may drain into a single confluence on a river network.

Since HydroJunctions are topologically linked to HydroEdges in the hydro network, the combination of this network and the other relationships shown in the analysis diagram means that the classes in the Arc Hydro framework are connected into an integrated data structure. This supports tracing water movement from one feature to another through the landscape. The creation of an integrated database, instead of a collection of data layers, is a key accomplishment of the Arc Hydro design in ArcGIS, providing a stronger foundation for building water resources applications in GIS than has previously existed.

So, how do you apply the Arc Hydro framework data model? All you need to begin is a set of streams, watersheds, water bodies, and points such as stream gage locations. In ArcCatalog™, you create a new, empty geodatabase containing a feature data set called Arc Hydro, copy your data layers into it, then use ArcCatalog's geometric network wizard to build a hydro network. You need to do this in ArcInfo since geometric networks cannot be built using ArcView. Then, you add the standard Arc Hydro attributes to the data using the Schema Creation Wizard in

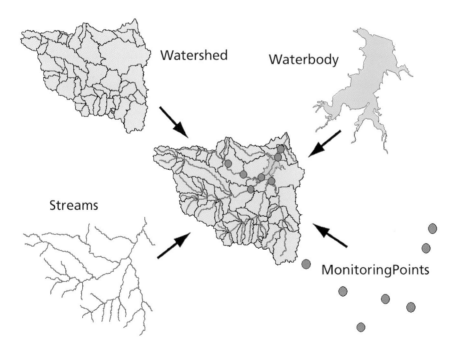

The Arc Hydro framework is a simplified version of the full Arc Hydro data model designed for an entry level user who just wants to put together a basic data set for streams, watersheds, water bodies, and locations representing stream gages and water-quality monitoring points.

ArcCatalog, and fill in the values of these attributes using the Arc Hydro tools. If you are beginning your study using a digital elevation model rather than an existing set of streams and watersheds, the Arc Hydro raster toolset can be used to generate the watersheds and a DEM-based stream network for inclusion in the framework. An important step is labeling all your features uniquely using the Assign HydroID tool. Once you've done that, you can connect the watersheds, water bodies, and gages to the network with other Arc Hydro tools, and you'll end up with an Arc Hydro framework data set that is simple enough to be created easily but powerful enough to support a significant range of applications. A tutorial on the Arc Hydro CD–ROM at the back of this book explains this process in more detail.

Once you've established the Arc Hydro framework data set, you can add additional Arc Hydro components, such as a 3-D river channel description and time series of water resources measurements.

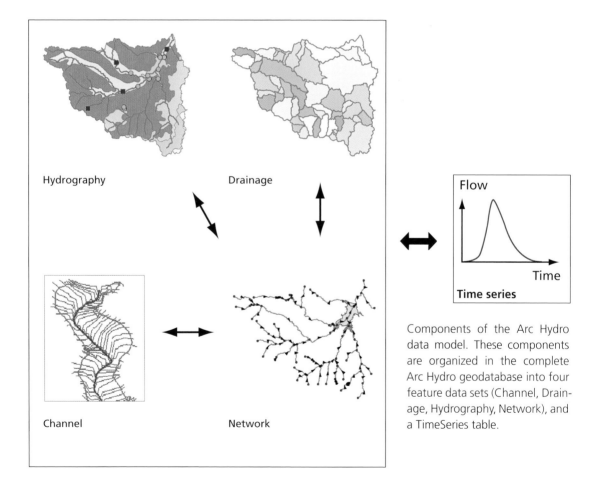

Hydrography

Drainage

Flow

Time

Time series

Channel

Network

Components of the Arc Hydro data model. These components are organized in the complete Arc Hydro geodatabase into four feature data sets (Channel, Drainage, Hydrography, Network), and a TimeSeries table.

Many core features of ArcGIS can be used immediately on an Arc Hydro framework data set, including the ArcGIS network-tracing tools. Arc Hydro provides an elegant tracing capability—the ability to trace upstream or downstream across drainage areas, so you can determine the region of hydrologic influence of any location on the landscape. Also, you can determine watershed properties, such as average precipitation or runoff, and accumulate them going downstream so that you can estimate the flow in streams and rivers. As the use of Arc Hydro continues to grow, further hydrologic analysis and modeling capabilities will become available. Yet, what is available now is practical and effective and is ready to be implemented in your water resources projects. The rest of the book provides conceptual and practical information about using Arc Hydro in water resources projects.

Arc Hydro data model components

The complete Arc Hydro data model divides water resources data into five components:
- Network—Connected sets of points and lines showing pathways of water flow
- Drainage—Drainage areas and stream lines defined from surface topography
- Channel—A 3-D line representation of the shape of river and stream channels
- Hydrography—The base data from topographic maps and tabular data inventories
- Time series—Tabular attribute data describing time-varying water properties for any hydro feature

The five components of Arc Hydro are described more fully in chapters 3 through 7.

Arc Hydro geodatabase structure

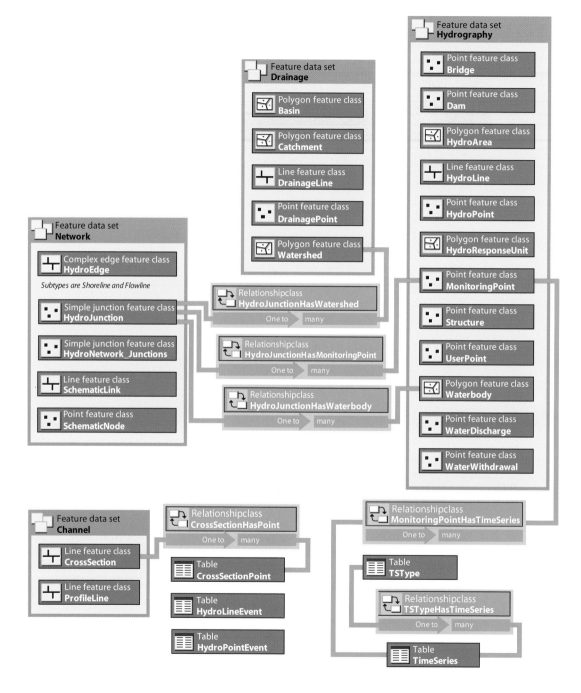

Hydro networks

Francisco Olivera, Texas A&M University
David Maidment, University of Texas at Austin
Dale Honeycutt, ESRI

The hydro network is the backbone of Arc Hydro, created from edges and junctions. The topological connection of its HydroEdges and HydroJunctions in a geometric network enables tracing of water movement upstream and downstream through streams, rivers, and water bodies. Relationships built from the HydroJunctions connect drainage areas and point features such as streamgaging stations to the hydro network. Locations on the hydro network are defined by a river-addressing scheme that defines where points are located on lines within drainage areas, allowing measurement of flow distance between any two points on a flow path.

Water flow networks

Networks of rivers and streams have always been fundamental to maps of the landscape, from the earliest maps to today's sophisticated GIS data layers. Traditionally, networks in GIS have been used to describe transportation systems of road networks. Such networks possess complications not seen in hydrology, such as flow moving in both directions along a road. Water flow in hydrology is driven by gravity and proceeds from higher to lower elevations. There are some exceptions to this rule, such as in tidal estuaries where the direction of flow can be reversible, but for nearly all cases of practical interest, flow in hydrologic networks moves in one direction, downhill.

A hydro network is a simplified representation of the blue lines on maps defining streams, rivers, and water bodies, in which centerlines can be drawn through all areal features to create a continuous, single-line network throughout the river system. River addressing, also called linear referencing, can be used to locate objects on the network.

Fluid flow can be classified as one-dimensional, two-dimensional, or three-dimensional, depending on how velocity changes with location in the coordinate system. Commonly, flow in streams and rivers is approximated as one-dimensional in the direction of motion, that is, in the direction parallel to the banks of the stream or river and transverse to stream cross sections. Thus, the distance between two points on a river is not measured between their x,y locations, but rather in terms of flow distance along the stream or river. A flow line traces the main direction of water movement in a one-dimensional flow. Flow properties such as discharge, velocity, depth, and constituent concentrations are allowed to vary only along the flow line, hence the term one-dimensional flow. A key virtue of the flow line is that it remains the same regardless of the size of the stream or river. A one-dimensional flow analysis for a large river like the Mississippi is carried out along its flow line in exactly the same way that a comparable analysis is performed for a small urban stream. A virtue of the assumption of one-dimensional flow is that the corresponding flow equations are readily solved for a wide range of cases. It is more difficult to solve the governing equations of two- and three-dimensional surface water flow systems.

When water flows into a water body such as a lake, its motion is no longer one-dimensional. However, to a degree of approximation, many lakes can also be considered as one-dimensional

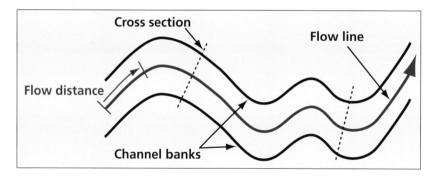

Flow lines trace the path of water movement in one-dimensional flow.

flow systems, where the flow tends to follow the channels of the inundated river. Hence, flow lines can be drawn through lakes from the point where a tributary inflow enters the lake to the point of discharge at the lake outlet. The process of drawing centerlines through water bodies can be automated using raster processing of areal features, or the centerlines can be digitized in ArcMap™ software.

Hydro networks

A hydro network is an ESRI geometric network made up of HydroEdges and HydroJunctions. As described in chapter 2, HydroEdges are built from the ArcObject complex edge features while HydroJunctions are built from ArcObject simple junction features. A geometric network needs junctions at both ends of every edge, and if a HydroJunction is not already located there, the ArcGIS geometric network builder creates a generic junction in a feature class called HydroNetwork_Junction. Thus, strictly speaking, a hydro network links three feature classes: HydroEdge, HydroJunction, and HydroNetwork_Junction.

During the design of Arc Hydro, many alternatives were examined to determine how best to construct the hydro network. The ArcGIS geometric network builder allows any number of line and point feature classes to participate in a network, subject to the rule that there cannot be two junctions sharing the same geographic location. For both network edges and junctions, we found it best to have a single feature class representing all feature types in the hydro network. So, for example, a junction representing a dam and another representing a stream gage are both represented as HydroJunctions. If necessary, the attribute FType can be used to differentiate among the types of HydroJunctions in a network.

In an early prototype hydro network, all the junctions were made into HydroJunctions, thereby eliminating the need for the generic HydroNetwork_Junctions. This alternative was discarded because HydroJunctions carry a large set of attributes requiring significant processing time to determine their values. Also, even though the HydroNetwork_Junction class then contained no features, the geometric network builder still created an empty HydroNetwork_Junction class as part of its operation. Considering all these factors, we recommend placing HydroJunctions at all locations of hydrologic interest on the network, and letting the geometric network builder create generic junctions wherever they are needed to complete the network topology.

A similar debate occurred about the definition of HydroEdge. In ArcGIS, when a feature class contains groups of features of distinctly different kinds, the feature class can be subtyped. This means that key attribute values can differ from one subtype to another, while in all other respects the features are treated the same as one another. HydroEdges are divided into two subtypes: flowlines (EdgeType = 1) that trace water movement through streams, rivers, and water bodies; and shorelines (EdgeType = 2) that form the interface between land and water for water bodies. Shorelines include those of lakes and reservoirs, coastlines to the sea or ocean, and bank lines for wide streams or rivers that are considered areal or water body features.

In early prototype versions of Arc Hydro, many more subtypes of HydroEdges were used, such as subtypes for natural channels, constructed channels, and pipelines. However, after examining many example applications, it became clear that trying to define an exhaustive list

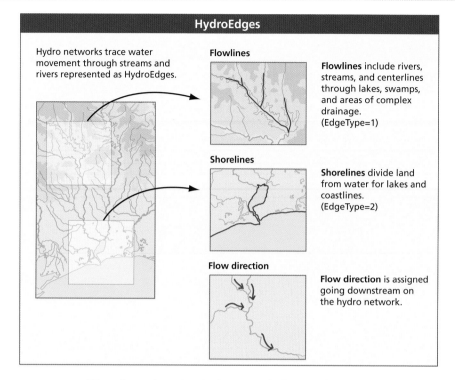

HydroEdges

Hydro networks trace water movement through streams and rivers represented as HydroEdges.

Flowlines

Flowlines include rivers, streams, and centerlines through lakes, swamps, and areas of complex drainage.
(EdgeType=1)

Shorelines

Shorelines divide land from water for lakes and coastlines.
(EdgeType=2)

Flow direction

Flow direction is assigned going downstream on the hydro network.

Water flows through a hydro network along HydroEdges.

The Netherlands as a hydro network. Flowlines are blue while Shorelines are red.

of potential subtypes of HydroEdges is an unending task because a data model designer can never anticipate all the variants a user may require. For this reason, the decision was made to have just two subtypes to distinguish flowlines through the center of the flowing water from shorelines around the borders of water bodies. For practical purposes, the most frequent use of the shoreline edge type is to identify coastlines. The shorelines of inland water bodies are readily depicted by the boundaries of water body polygons. If necessary, such as for a cartographic display distinguishing ephemeral streams from perennial streams, the user can further distinguish among the different types of HydroEdges using the FType attribute, the same as for HydroJunctions.

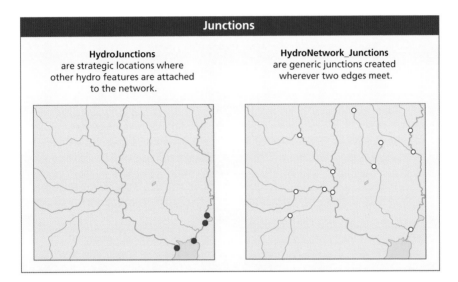

HydroJunctions and HydroNetwork_Junctions are points on the hydro network.

Flow direction

As part of geometric network construction, the direction of flow is assigned to each HydroEdge in order to direct flow toward the nearest sink. A HydroJunction is a sink if it acts as an outlet for discharge of water from the network, such as where a river system discharges into the ocean. To identify flow direction, the hydrologist first identifies those HydroJunctions that serve as sinks (assigning them that task through their AncillaryRole attribute), then uses ArcGIS Network Analyst tools to determine flow direction on each edge. For most hydro networks, this automated process successfully assigns the correct flow direction to most HydroEdges. Wherever there are loops in the network, however, flow direction cannot be assigned automatically because the downstream direction cannot be determined from graphical information alone. Loops occur in braided stream channels and in networks of constructed channels, such as irrigation canal systems. When this occurs, the flow direction must be assigned manually.

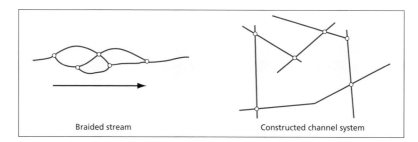

Braided stream   Constructed channel system

The HydroEdge flow direction attribute, FlowDir, is defined by a Coded Value Domain where each number has an associated text label. This domain possesses four values: 0 (uninitialized), 1 (with digitized), 2 (against digitized), and 3 (indeterminate). The default value is 0. "With digitized" and "against digitized" compare the flow direction to the direction in which the line was digitized, as indicated by the order of the vertices defining each segment of the line. ArcGIS Network Analyst symbolizes flow direction on network edges with flow direction values of 1 or 2 using an arrow, while edges that have no assigned flow direction are symbolized with dots.

Before the advent of ArcGIS, the simplest way to ensure that all network edges had an unambiguous flow direction was to artificially break all loops using manual editing of the network. This practice distorts the map cartography, creates artificial breaks in flow paths that do not exist in nature, and is no longer necessary. All ArcGIS network edges carry an attribute Enabled, defined by a Coded Value Domain whose values are either disabled (0) or enabled (1) with a default value of Enabled. By setting this attribute value to disabled, the flow in an edge is automatically blocked, which is also symbolized by a dot in the network diagram.

The hydro network is a highly useful data structure for supporting hydrologic analysis. During the three-year design process for Arc Hydro, researchers at the Center for Research in

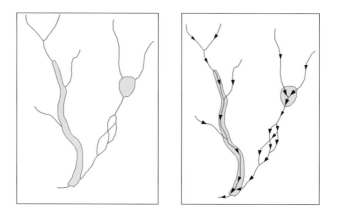

Transition from map hydrography to network hydrography. Flow direction on HydroEdges is symbolized by arrows.

Water Resources built stream networks for each of the 23 river and coastal drainage basins of Texas. This experience convinced us that Arc Hydro users needed more control over flow direction on HydroEdges than is available in the automated flow direction assignment process using network analysis tools in ArcGIS. Accordingly, the Arc Hydro toolset contains an Assign Flow Direction tool that allows for manual editing of flow direction, and the storage and retrieval of the assigned flow direction using the Arc Hydro attribute FlowDir. Using this tool, hydrologists can carefully examine the topography surrounding the network, and manually assign a flow direction in looped drainage systems. Coastal regions are often very flat and their drainage patterns are dominated by constructed flow channels, so manual flow direction assignment is needed for most HydroEdges in these areas. The hydro network for the Netherlands is an example of such a system.

The process of taking a set of blue lines from digitized mapped streams and constructing a hydro network from them is lengthy and iterative. The hydrologist first has to ensure that all the hydrographic lines that should be connected are actually connected, then that the flow direction on each edge has been assigned appropriately. Once the hydro network is complete, however, it is a powerful instrument for supporting hydrologic analysis.

Connecting features to HydroJunctions

The topology of the hydro network connects only points and lines. However, areas play an important role in hydrology as well, so a way to connect them to the network is also needed. Two types of areas are particularly important: water bodies and drainage areas.

During the Arc Hydro design process, several schemes for connecting water bodies to the network were examined. For example, an early variant required that all water bodies have their shorelines embedded in the hydro network, but this greatly complicated assigning flow direction to the HydroEdges, and was unnecessary because shorelines do not carry flow. In the end, it was decided that the simplest way to connect water bodies to the network was to construct a relationship between each Waterbody feature and the corresponding HydroJunction located at its downstream outlet. This is done by assigning the value of the HydroID of the outlet HydroJunction as the value of the JunctionID attribute of the Waterbody (an example of this is presented at the end of this chapter). The Arc Hydro toolset contains an Assign Area Outlets tool that carries out this assignment automatically.

Drainage areas are similarly attached to the hydro network by assigning the HydroID of the HydroJunction at the drainage area outlet location as the JunctionID attribute of the appropriate drainage area feature class (Catchment, Watershed, or Basin). This topological connection of "areas flow to points on lines" is particularly appropriate for hydrology, because it enables tracing of the movement of a raindrop from a drainage area onto a HydroJunction, and thence downstream on the hydro network. Any number of drainage areas can use the same HydroJunction as their outlet location, which creates a simple, flexible way of linking drainage from the land surface to flow in water systems. This subject is explored more fully in chapter 4.

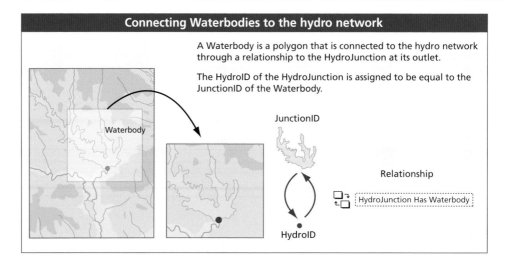

Connecting Waterbodies to the hydro network

A Waterbody is a polygon that is connected to the hydro network through a relationship to the HydroJunction at its outlet.

The HydroID of the HydroJunction is assigned to be equal to the JunctionID of the Waterbody.

JunctionID

Relationship

HydroJunction Has Waterbody

HydroID

Once it was recognized during the Arc Hydro design process that the HydroID-to-JunctionID relationship connected drainage areas and water bodies to the hydro network, it was obvious that other features should be similarly related to the network. In particular, point features such as MonitoringPoints representing stream gages or water quality sampling sites can be related to the network by creating a HydroJunction at the nearest point on the network edge to the point feature, then assigning the HydroID of the HydroJunction as the JunctionID of the Monitoring-Point. This process has the advantage of leaving point features at their true x,y location and not forcing them to be located exactly on the network line (only the related junction must be on the network). For example, a point feature symbolizing a stream gage should be located on the bank of the stream at the position of the gage house containing the recording equipment. It has

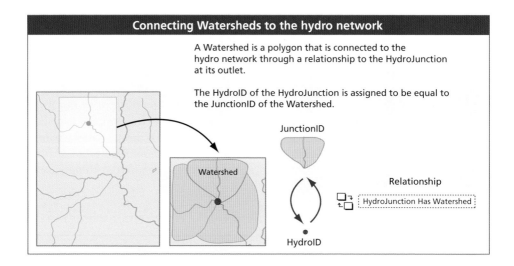

Connecting Watersheds to the hydro network

A Watershed is a polygon that is connected to the hydro network through a relationship to the HydroJunction at its outlet.

The HydroID of the HydroJunction is assigned to be equal to the JunctionID of the Watershed.

JunctionID

Relationship

HydroJunction Has Watershed

HydroID

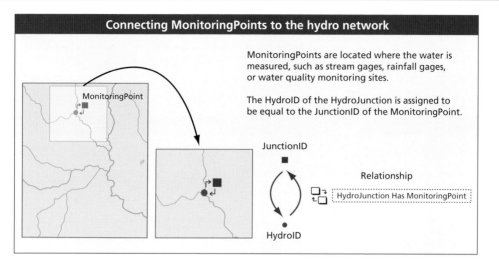

Connecting MonitoringPoints to the hydro network

MonitoringPoints are located where the water is measured, such as stream gages, rainfall gages, or water quality monitoring sites.

The HydroID of the HydroJunction is assigned to be equal to the JunctionID of the MonitoringPoint.

JunctionID

Relationship

HydroJunction Has MonitoringPoint

HydroID

been a common practice when associating point features with a network to snap them onto the network lines, and in effect to move the gage house into the center of the stream! Using the HydroID–JunctionID relationship avoids this problem. These HydroJunctions are marker points on the hydro network used to signal the presence of adjacent point features.

The Arc Hydro framework data model presented in chapter 2 contains three relationships between HydroJunction features and Waterbody, Watershed, and MonitoringPoint features, respectively. All these relationships use the HydroID-to-JunctionID association just described. For the same reasons, the JunctionID attribute is included on all point feature classes in the hydrography feature data set, such as WaterWithdrawal and WaterDischarge points. A formal relationship is not constructed between all of these feature classes and the HydroJunction class because that would unnecessarily clutter the data model with relationships that may not always be needed. Wherever necessary, you can add a relationship to Arc Hydro using ArcCatalog.

Hydro navigation

Hydro navigation is a process of tracing water movement from feature to feature through the landscape. Arc Hydro supports three types of navigation:

- Network navigation—Carried out using the ArcGIS network tracing tools applied to the hydro network.
- Next downstream navigation—Carried out on HydroJunctions or other feature classes that have a NextDownID attribute assigned to them.
- Schematic navigation—Carried out on a pair of separate feature classes, called SchematicLink and SchematicNode, that serve to link strategic locations with straight lines in a schematic diagram.

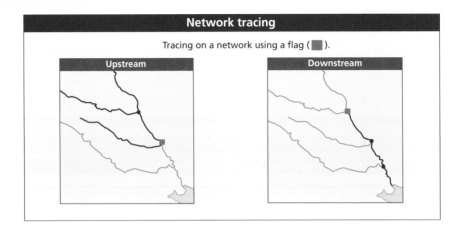

Network tracing

Tracing on a network using a flag (▪).

Upstream

Downstream

Network navigation

Network navigation traces the movement of water or pollutant particles through the network edges and junctions. Using network analysis tools in ArcGIS, all tracing functions are initiated by placing a flag at the network location from which the trace originates. Upstream tracing identifies all edges and junctions that drain to the flag location. By invoking the relationship between drainage areas and the selected HydroJunctions, the upstream drainage area is readily identified and may be used to identify the potential areas where a water or pollutant particle entered the flow system. Downstream tracing, on the other hand, identifies all edges and junctions that constitute the flow path from the flag location to the network outlet or sink. For example, network navigation can be used to identify all the downstream features affected by a particular wastewater discharge point.

Next downstream navigation

During the process of network navigation it is easy to identify which HydroJunction is the next one downstream of a given HydroJunction. The HydroID of the next downstream HydroJunction is assigned as the NextDownID of the upstream HydroJunction. This process works in dendritic networks, that is, in networks where each HydroJunction has one and only one downstream HydroJunction, and there are no cases where water cycles around through a sequence of Hydro-Junctions and ends up back at its original location. Flow driven downhill by the earth's gravity force cannot cycle back to its original point—a pumped or tidal flow system is required for this to happen.

Tracing using NextDownID occurs only within a single feature class, in this case, HydroJunctions. Tracing proceeds from a selected feature to identify a set of selected features, located either upstream or downstream. This capability is useful for answering queries like: find the next downstream stream gage so a flow estimate can be made, or find all the upstream water withdrawal points whose withdrawals may affect this withdrawal point. An extension of the same concept for tracing across drainage areas using the Next Downstream navigation tool is described in chapter 4.

Tracing using NextDownID does not require that all the features be graphically connected using network lines, just that each feature know what is the next downstream feature. This is a simpler method of tracing than network navigation and lends itself to accumulating attributes going downstream through a set of features, such as accumulating the area of all upstream watersheds to determine the drainage area at each HydroJunction. The Arc Hydro toolset contains tools for next downstream navigation and attribute accumulation to automate these processes.

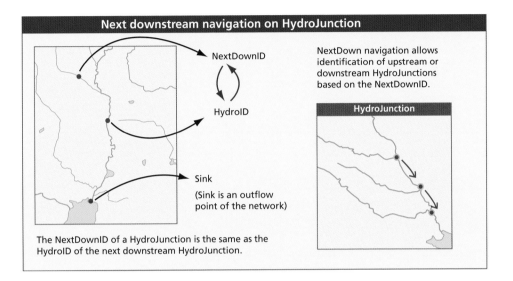

Next downstream navigation on HydroJunction

NextDownID

HydroID

NextDown navigation allows identification of upstream or downstream HydroJunctions based on the NextDownID.

HydroJunction

Sink

(Sink is an outflow point of the network)

The NextDownID of a HydroJunction is the same as the HydroID of the next downstream HydroJunction.

Schematic navigation

A schematic network is a simplification of the hydro network that consists of separate point and line feature classes called SchematicNode and SchematicLink, respectively. The schematic network is an abstract representation of the elements to which the hydrologic models are applied, and it provides a simplified view of the connectivity of those elements in the landscape. SchematicLinks are attributed with FromNodeIDs and ToNodeIDs that refer to the HydroIDs of the SchematicNodes at their end points. Each SchematicNode carries a FeatureID attribute that refers to the underlying hydro feature that the node represents. The SchematicLink and SchematicNode classes are built using regular point and line feature classes in order to facilitate the construction of schematic networks. These networks can even be constructed with ArcView, which does not have the capability of building a geometric network. Schematic networks are useful as a visual check to make sure that the hydrologic elements needed for a model are correctly linked in the landscape. If formal network tracing is required on a schematic network, its point and line features can be built into a geometric network in ArcCatalog.

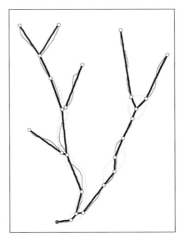

A schematic network is comprised of SchematicLinks and SchematicNodes.
SchematicLinks are represented by black lines and SchematicNodes by white dots.

River addressing

River addressing is used to locate objects on a river or stream system. These addresses are constructed using the linear referencing tools in ArcGIS. Traditionally in hydrology, river addressing has been done using river miles or kilometers measured upstream from the river mouth.

Lock and Dam 5a is located on Mississippi River mile 728.5 near Fountain City, Wisconsin.
River mile 728.5 is measured upstream from the confluence of the Mississippi and Ohio Rivers.

All the HydroEdges and HydroJunctions in Arc Hydro carry the attribute LengthDown, which is the distance to the nearest network sink, usually calculated in kilometers based on the HydroEdge attribute LengthKm. The Arc Hydro toolset contains a tool that automatically populates the LengthDown attribute. LengthDown is a useful attribute because it assigns a measure of flow distance to the outlet everywhere within the network. For two locations on a flow path, the flow distance between them is the difference between their LengthDown values.

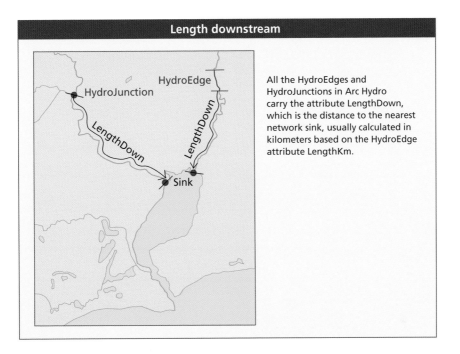

Length downstream

All the HydroEdges and HydroJunctions in Arc Hydro carry the attribute LengthDown, which is the distance to the nearest network sink, usually calculated in kilometers based on the HydroEdge attribute LengthKm.

In ArcGIS, measure, or m-values, can be assigned to every vertex along a line, so that each vertex knows where it lies in terms of flow distance. Measures can be assigned to vertices using absolute addressing, which consists of using absolute distances from an arbitrarily defined origin, or relative addressing, which consists of using the relative distance along a HydroEdge or a river reach.

In hydraulic modeling of river channels, it is customary to assign measures (called river stationing in the United States and river chainage in Europe) that specify the flow distance in meters or feet between a point on the channel network and a reference point located either at the upstream or downstream end of the river reach being modeled. This is an example of absolute addressing.

The HydroEdge attribute ReachCode identifies a set of HydroEdges linearly connected to form a single river reach, usually defined between stream confluences. The U.S. National Hydrography Dataset labels all the river reaches of the United States with a unique ReachCode that is a 14-digit number whose first eight digits specify which Hydrologic Unit Code subbasin

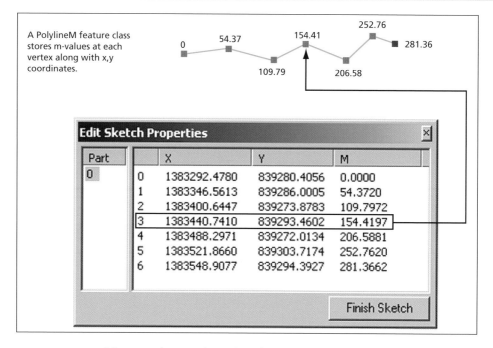

Measure values can be assigned to each vertex on a line.

the reach lies within, and whose final six digits identify the river segment within the subbasin. Information is linearly referenced on the U.S. National Hydrography Dataset using relative addressing where the zero percent location is at the downstream end of each reach, and the 100 percent location is at the upstream end of the reach. The result is somewhat similar to a street address. For example, 123 Oak Avenue, Austin, TX 78758, gives a point location (123) on a line (Oak Avenue) within an area (the ZIP Code 78758). The analogous process for a river address on the U.S. National Hydrography Dataset is a percent location on a stream segment within a particular Hydrologic Unit Code subbasin.

Relative addressing's advantage is that, because relative distances along lines barely change with the scale of the data, points can be reliably located regardless of the map scale of the under-lying stream network data set. During the development over several decades of river reach files for the United States, it became awkward to use absolute addressing when hydrographic lines from larger-scale maps (e.g., 1:24,000 scale) replaced those previously used from smaller-scale maps (e.g., 1:100,000 scale). Larger-scale maps have more crenelated flow lines and longer flow distances than do smaller-scale maps, so all the river mile or kilometer addresses change when the map scale changes, or even when river channels are modified, such as when hydrologists do river straightening. Relative addressing overcomes these problems to some degree and pro-duces a greater degree of permanence of river address values.

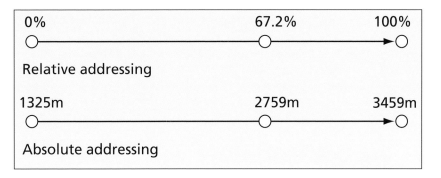

Measures are assigned using a relative or an absolute addressing system.

Hydro events

Event tables in ArcGIS are points or lines that are not explicitly defined with their own geographic coordinates, but are instead referenced to a separate line feature class on which an addressing system has been defined. This is analogous to specifying that a house is located at 123 Oak Avenue, rather than giving its latitude and longitude.

Objects located by linear referencing on a hydro network are called hydro events, which may be HydroPointEvents (i.e., a point location on a HydroEdge) or HydroLineEvents (i.e., a line between two identified points on a HydroEdge). Hydro events carry a ReachCode attribute to identify which HydroEdge or set of HydroEdges they are referenced to. Additionally, HydroPointEvents have a Measure attribute to specify where within that reach the point is located, while HydroLineEvents have FMeasure and TMeasure attributes, to locate the endpoints of the linear event on the reach. Line events also carry the attribute Offset, which allows line events to be displaced to the left or right of the reference line, such as when separate information is defined for the left bank of a river as compared to the right bank.

Once events are defined, any number of attributes can be added to them. Events are an alternative way of relating information to the hydro network that is a lot less burdensome to the network than putting junctions everywhere there is a hydrologic feature of interest. If necessary, events can be converted to point or line features using the event display tools in ArcGIS. Linear referencing of events using the ReachCode is how the U.S. Environmental Protection Agency locates information on the U.S. National Hydrography Dataset, such as information about wastewater discharge points along a river.

Hydro network for the Guadalupe basin

The CD–ROM at the back of the book contains an Arc Hydro framework data set for the Guadalupe Basin. This basin in south Texas covers a drainage area of 15,500 square kilometers. The river carries flow from the dry Texas hill country in the upper portions of the basin through the flat coastal plain in the lower portions to discharge into the Gulf of Mexico. The U.S. Geological Survey has subdivided the basin into four Hydrologic Unit Code subbasins that are stored in the Arc Hydro feature class Watershed. A total of 2,860 HydroEdges were built from the U.S. National Hydrography Dataset at a scale of 1:100,000, to which were joined 235 water bodies selected from the U.S. National Hydrography Dataset as having a surface area of more than 0.6 square kilometers. The data set contains 29 MonitoringPoints, which are U.S. Geological Survey streamflow-gaging stations. HydroJunctions are used to relationally connect all the watersheds, water bodies, and monitoring points to the hydro network.

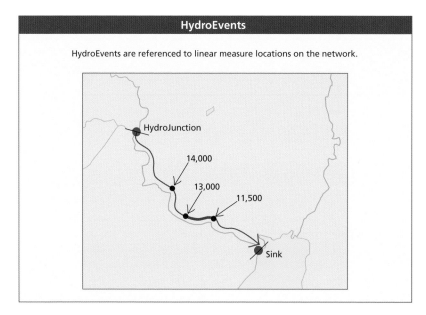

A HydroPointEvent is located at measure 14,000. A HydroLineEvent is located between measures 13,000 and 11,500.

Canyon Lake is a large controlled reservoir located at the outlet of the Upper Guadalupe subbasin. Releases of water from Canyon Lake spill fresh, cold water into the Guadalupe River. Downstream of Canyon Lake, riding inflated tubes down the cool reach of the Guadalupe River is a popular summertime activity.

The following illustrations show how the main concepts in this chapter are applied in the Guadalupe basin, with a particular focus on Canyon Lake.

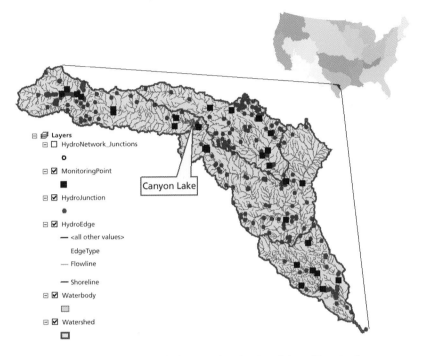

Arc Hydro framework data set for the Guadalupe River basin

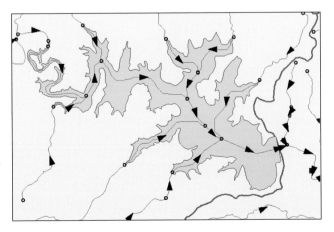

The hydro network carrying flow through Canyon Lake. The arrowheads show flow direction.

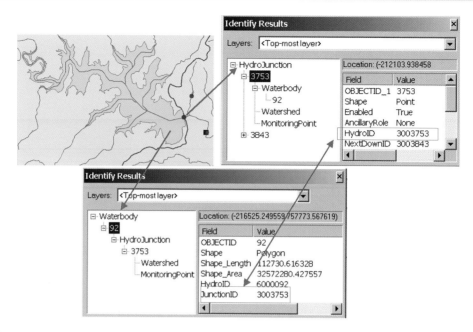

The HydroJunction with HydroID 3003753 at the Canyon Lake outlet is relationally connected to the Waterbody feature by storing 3003753 as the JunctionID of the Waterbody. Similar relationships connect Watersheds and MonitoringPoints to HydroJunctions.

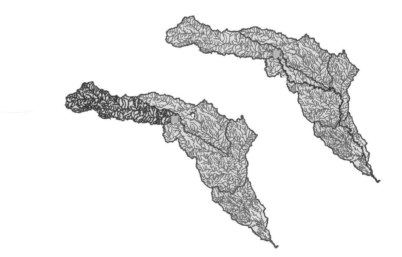

Upstream and downstream traces on the hydro network from Canyon Lake. The green square symbolizes a network flag that identifies the origin of the trace. All HydroEdges and HydroJunctions on the traced paths are selected.

50

Data dictionary

These diagrams summarize the object and feature classes in the hydro network component of Arc Hydro, and their interrelationships. All the classes shown are available for loading data, because they have inherited all the attributes from classes located above them in the UML hierarchy. The attributes shaded in blue are ESRI standard attributes, while those shaded in white are Arc Hydro attributes. The terms used here are defined in the glossary at the back of this book.

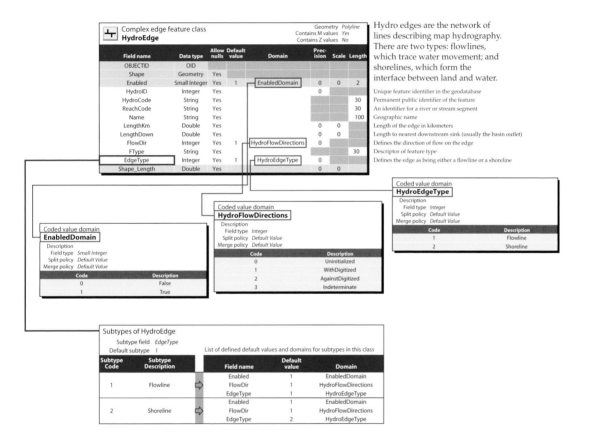

Hydro edges are the network of lines describing map hydrography. There are two types: flowlines, which trace water movement; and shorelines, which form the interface between land and water.

51

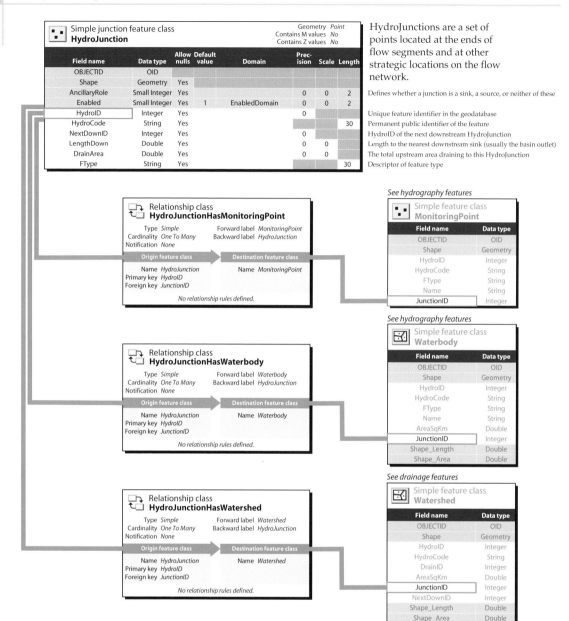

Simple junction feature class
HydroJunction

Geometry *Point*
Contains M values *No*
Contains Z values *No*

Field name	Data type	Allow nulls	Default value	Domain	Prec-ision	Scale	Length
OBJECTID	OID						
Shape	Geometry	Yes					
AncillaryRole	Small Integer	Yes			0	0	2
Enabled	Small Integer	Yes	1	EnabledDomain	0	0	2
HydroID	Integer	Yes			0		
HydroCode	String	Yes					30
NextDownID	Integer	Yes			0		
LengthDown	Double	Yes			0	0	
DrainArea	Double	Yes			0	0	
FType	String	Yes					30

HydroJunctions are a set of points located at the ends of flow segments and at other strategic locations on the flow network.

Defines whether a junction is a sink, a source, or neither of these

Unique feature identifier in the geodatabase
Permanent public identifier of the feature
HydroID of the next downstream HydroJunction
Length to the nearest downstream sink (usually the basin outlet)
The total upstream area draining to this HydroJunction
Descriptor of feature type

Relationship class
HydroJunctionHasMonitoringPoint

Type *Simple*
Cardinality *One To Many*
Notification *None*

Forward label *MonitoringPoint*
Backward label *HydroJunction*

Origin feature class
Name *HydroJunction*
Primary key *HydroID*
Foreign key *JunctionID*

Destination feature class
Name *MonitoringPoint*

No relationship rules defined.

See hydrography features

Simple feature class
MonitoringPoint

Field name	Data type
OBJECTID	OID
Shape	Geometry
HydroID	Integer
HydroCode	String
FType	String
Name	String
JunctionID	Integer

Relationship class
HydroJunctionHasWaterbody

Type *Simple*
Cardinality *One To Many*
Notification *None*

Forward label *Waterbody*
Backward label *HydroJunction*

Origin feature class
Name *HydroJunction*
Primary key *HydroID*
Foreign key *JunctionID*

Destination feature class
Name *Waterbody*

No relationship rules defined.

See hydrography features

Simple feature class
Waterbody

Field name	Data type
OBJECTID	OID
Shape	Geometry
HydroID	Integer
HydroCode	String
FType	String
Name	String
AreaSqKm	Double
JunctionID	Integer
Shape_Length	Double
Shape_Area	Double

Relationship class
HydroJunctionHasWatershed

Type *Simple*
Cardinality *One To Many*
Notification *None*

Forward label *Watershed*
Backward label *HydroJunction*

Origin feature class
Name *HydroJunction*
Primary key *HydroID*
Foreign key *JunctionID*

Destination feature class
Name *Watershed*

No relationship rules defined.

See drainage features

Simple feature class
Watershed

Field name	Data type
OBJECTID	OID
Shape	Geometry
HydroID	Integer
HydroCode	String
DrainID	Integer
AreaSqKm	Double
JunctionID	Integer
NextDownID	Integer
Shape_Length	Double
Shape_Area	Double

Simple junction feature class
HydroNetwork_Junctions

Geometry *Point*
Contains M values *No*
Contains Z values *No*

Field name	Data type	Allow nulls	Default value	Domain	Precision	Scale	Length
OBJECTID	OID						
SHAPE	Geometry	Yes					
Enabled	Small Integer	Yes	1	EnabledDomain	0	0	2

Generic junctions created during the building of a geometric network at the ends of all the edges, except where a HydroJunction exists there

Simple feature class
SchematicNode

Geometry *Point*
Contains M values *No*
Contains Z values *No*

Field name	Data type	Allow nulls	Default value	Domain	Precision	Scale	Length
OBJECTID	OID						
Shape	Geometry	Yes					
HydroID	Integer	Yes			0		
HydroCode	String	Yes					30
FeatureID	Integer	Yes			0		

A representative point for a hydro feature connected into a schematic network

Unique feature identifier in the geodatabase
Permanent public identifier of the feature
HydroID of the associated hydro feature

Simple feature class
SchematicLink

Geometry *Polyline*
Contains M values *No*
Contains Z values *No*

Field name	Data type	Allow nulls	Default value	Domain	Precision	Scale	Length
OBJECTID	OID						
Shape	Geometry	Yes					
HydroID	Integer	Yes			0		
HydroCode	String	Yes					30
FromNodeID	Integer	Yes			0		
ToNodeID	Integer	Yes			0		
Shape_Length	Double	Yes			0	0	

A straight line connecting two Schematic Nodes in a schematic network

Unique feature identifier in the geodatabase
Permanent public identifier of the feature
HydroID of the schematic node at the from end of the link
HydroID of the schematic node at the to end of the link

Table
HydroLineEvent

Field name	Data type	Allow nulls	Default value	Domain	Precision	Scale	Length
OBJECTID	OID						
ReachCode	String	Yes					30
FMeasure	Double	Yes			0	0	
TMeasure	Double	Yes			0	0	
Offset	Double	Yes			0	0	

An attribute or set of attributes associated with a line segment through measure values

An identifier for a river or stream segment
Measure value at the start of the line event
Measure value at the end of the line event
Distance from the center of the line to display event

Table
HydroPointEvent

Field name	Data type	Allow nulls	Default value	Domain	Precision	Scale	Length
OBJECTID	OID						
ReachCode	String	Yes					30
Measure	Double	Yes			0	0	

An attribute or set of attributes associated with a location on a line segment through a measure value

An identifier for a river or stream segment
Measure value for the point event

53

Drainage systems

Francisco Olivera, Texas A&M University
Jordan Furnans, University of Texas at Austin
David Maidment, University of Texas at Austin
Dean Djokic, ESRI
Zichuan Ye, ESRI

Water and land interact with one another: the shape of the land surface directs the drainage of water through the landscape, while the erosive power of water slowly reshapes the land surface. Streams, rivers, and water bodies lie in the valleys and hollows of the land surface, and drainage from the ridges and higher land areas flows downhill into these water systems. Digital elevation models are used to analyze the drainage patterns of the land-surface terrain, and drainage areas are delineated from outlets chosen either manually or automatically according to physical rules. Raster analysis using fine-resolution DEMs is practical only over limited areas, but these results may be combined with vector networks to carry out regional hydrologic studies. Drainage areas can be traced upstream and downstream, either through their attachment to the hydro network or by using area-to-area navigation, thereby identifying the region of hydrologic influence upstream and downstream of a catchment or watershed.

Drainage from the landscape

Precipitation falls on the land surface, soaks into soils, evaporates, or runs off the land surface into streams. Drainage is the flow process by which water moves from the point that it falls onto the landscape down to a stream, then to a river, and finally to the sea. Drainage flows downhill so the wetness of the land surface is a function of its shape, ridges being higher and having dryer soils and valleys being lower and having wetter soils. The shape of the landscape directs the flow of water but is in turn shaped by the erosive power of water. Gradually, over geologic time, the familiar patterns of small drainage paths leading to ditches, then to streams, and finally to rivers are etched on the landscape by flowing water as it carries away eroded sediment. There is thus an intimate relationship between the shape of the land-surface terrain and the stream network that exists within it.

Land-surface terrain and stream network for the San Marcos basin.
The San Marcos River is a tributary of the Guadalupe River.

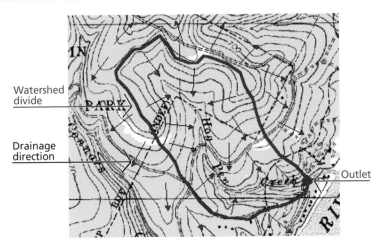

Watershed divide

Drainage direction

Outlet

Manual delineation of drainage area on a topographic map

Traditionally, drainage areas have been delineated from topographic maps, where drainage divides are located by analyzing the contour lines. Arrows representing water flow direction can be drawn perpendicular to each contour, in the direction of the steepest descent. The location of a drainage divide is where flow directions diverge. The drainage area boundary is digitized by drawing a continuous line transverse to the contours, up from the outlet point to the ridgeline, along the ridgeline around the drainage area, then descending to the outlet point again.

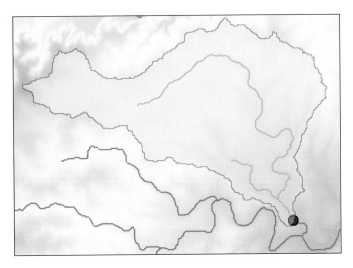

Drainage area delineation using a digital elevation model

Drainage areas can also be delineated automatically using digital elevation models of the land-surface terrain. A digital elevation model is a grid of square cells, where each cell value represents the elevation of the land surface. By determining how water flows from cell to cell, the set of cells whose drainage flows through the cell at the outlet point location can be identified, and thus the drainage area determined.

The Arc Hydro toolset contains functions to accomplish automated drainage area and stream network delineation from digital elevation models, and there are many advantages to using them. Drainage patterns are complex and automated processing removes the need for the hydrologist to spend endless hours trying to interpret drainage patterns from contour lines on maps. Nevertheless, digital elevation models are only an approximate representation of land-surface terrain and manual editing of drainage boundaries may be necessary, especially in regions with very flat terrain having many constructed rather than natural drainage channels.

The Arc Hydro data model accepts drainage areas and connects them to the hydro network, no matter whether the areas were automatically or manually delineated, and allows for the fact that the mapped stream hydrography may not be entirely consistent with the land-surface terrain or hypsography used to determine drainage area or boundaries.

Raindrop model

The Arc Hydro data model represents a system of behavior. In the most simplistic sense, this system follows the drainage of a raindrop from when it hits the ground through streams and rivers to the point when it comes to its final resting place, usually in the ocean. At the beginning of its journey, we could ask the question: "If a raindrop falls between two streams, to which stream does it flow?" Fortunately, when digital elevation models are used, the cell-to-cell flow path of a raindrop is easy to trace (the Arc Hydro toolset has a function for accomplishing this), and the flow paths of raindrops falling anywhere on the land surface can be identified.

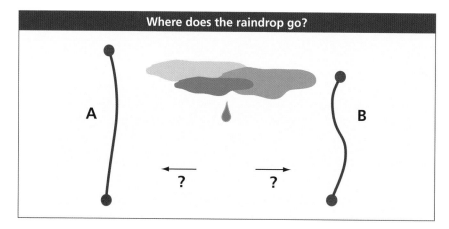

To which stream does the raindrop flow?

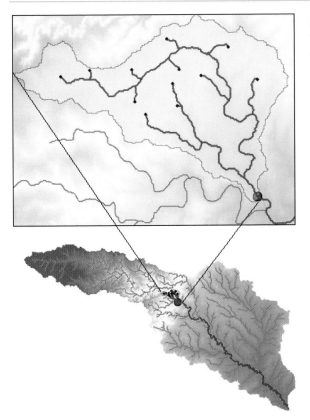

Following the paths of raindrops from where they fall on the land surface to streams, rivers, and all the way to the ocean.

The cell-to-cell flow paths on raster grids make it easy to define the flow path of a raindrop, but the issue is more complex when vector data is used to define drainage areas. Then, the drainage area polygon within which the raindrop falls needs to know where on the stream network this raindrop discharges. The only location that can be unambiguously identified is the drainage area outlet, that is, it can be stated certainly that the effect of a raindrop falling anywhere within a drainage area has been felt by the time the flow reaches the drainage area outlet. In Arc Hydro, the connection "areas flow to lines at points," is made and this forms the basis of building the relationship between Watershed and HydroJunction features.

It is possible, however, to associate drainage areas with streams more simply: each stream segment has one and only one drainage area associated with it. In that case, following the path of a raindrop from anywhere on the land surface to the stream network consists of identifying the stream segment lying within the drainage area where the raindrop fell. This simpler "area to line" connectivity of the land surface to the stream network is used for the Arc Hydro Catchment processing system, subdividing the landscape into a large number of elementary drainage areas, each with its own associated DrainageLine and DrainagePoint at the outlet.

Drainage areas

Drainage area boundaries are used in water-availability studies, water-quality projects, flood forecasting programs, as well as many other engineering and public policy applications. So far in this book, the terms drainage area, catchment, watershed, and basin have been used interchangeably without defining the distinctions between them. Indeed in normal language there is little distinction between these terms. In the United States, watershed is the standard term for a drainage area, while in Britain and Europe, catchment is considered the standard term. Basin is understood to be a larger drainage area associated with a major river, such as for the Mississippi or Missouri River basins.

In Arc Hydro, the following definitions apply:

- Basin—A set of administratively chosen drainage areas that partition a region for purposes of water management. Basins are usually named after the principal rivers and streams of a region. Basins may serve as spatial packaging units for Arc Hydro data sets.
- Watershed—A tessellation or subdivision of a basin into drainage areas selected for a particular hydrologic purpose. Watersheds may drain to points on a river network, to river segments, or to water bodies.
- Catchment—A tessellation or subdivision of a basin into elementary drainage areas defined by a consistent set of physical rules.

A distinction is drawn between catchments, whose layout can be automatically determined using a set of rules applied to a digital elevation model, and watersheds, whose outlets are chosen manually to serve a particular hydrologic purpose. Watershed outlet points may lie anywhere, and are not necessarily coincident with catchment outlet points. Drainage area is used in this book as a generic term for a catchment, watershed, or basin, and also to specify the physical area within the drainage basin boundary. The context in which the term is used serves to distinguish these two meanings of drainage area.

Catchments

Catchments are elementary drainage areas defined by a consistent set of physical rules. The approach most often used is to define the beginning of a stream using a threshold upstream drainage area, then delineate catchments from the confluences of all the stream segments thus created. The analysis is carried out on a digital elevation model, so a catchment is defined by a zone of raster cells, a stream segment is defined by a line of raster cells, and an outlet is defined by a single cell. When converted from raster to vector features, these cells form the Arc Hydro feature classes Catchment, DrainageLine, and DrainagePoint, respectively. The advantage of the threshold drainage area approach to defining catchments is that the tessellation of the landscape into elementary drainage areas is readily automated. The disadvantage is that this approach takes no account of the presence of water bodies in the landscape (they are treated as if they are large, flat land surfaces), and it cannot produce catchments draining directly to the coastline rather than to a stream.

An alternative approach to defining catchments is to take a hydro network with reach codes defined for all of its segments, and use a digital elevation model to delineate an elementary drainage area for each reach. This approach allows for catchments to be defined for coastline

Scales of representation of drainage systems

At the highest level are Basins, which may be subdivided into Watersheds or Catchments. Digital Elevation Models may be used to define drainage area boundaries for Catchments, Watersheds, and Basins.

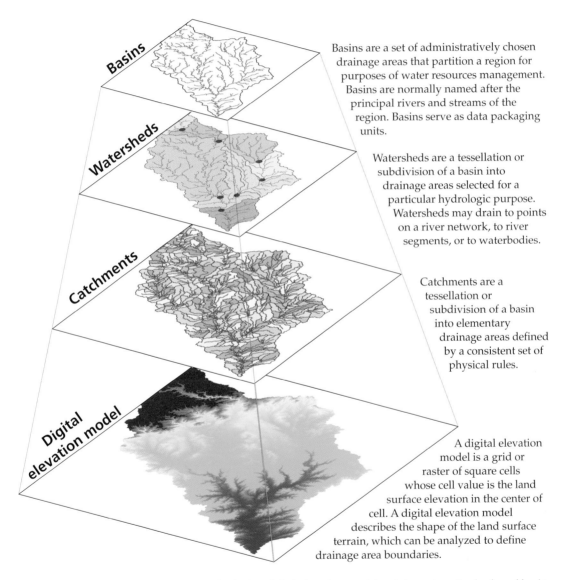

Basins

Basins are a set of administratively chosen drainage areas that partition a region for purposes of water resources management. Basins are normally named after the principal rivers and streams of the region. Basins serve as data packaging units.

Watersheds

Watersheds are a tessellation or subdivision of a basin into drainage areas selected for a particular hydrologic purpose. Watersheds may drain to points on a river network, to river segments, or to waterbodies.

Catchments

Catchments are a tessellation or subdivision of a basin into elementary drainage areas defined by a consistent set of physical rules.

Digital elevation model

A digital elevation model is a grid or raster of square cells whose cell value is the land surface elevation in the center of cell. A digital elevation model describes the shape of the land surface terrain, which can be analyzed to define drainage area boundaries.

Four spatial scales for representation of a basin: digital elevation model, catchment, watershed, and basin segments and for water bodies, as well as for river reaches, but it is more difficult to implement than the simple definition of catchments based on threshold drainage area.

Watersheds

How water resources agencies subdivide a given landscape into drainage units may vary significantly from one agency to another, or even within the same agency when different kinds of analyses are undertaken. In the United States, the National Weather Service forecasts floods on all the major rivers. For that purpose the nation is divided into watersheds, each watershed being a model unit in the agency's river forecast system. The Environmental Protection Agency manages water pollution using Total Maximum Daily Loads defined on watersheds draining to selected river segments or water bodies, a different watershed layout than that used by the National Weather Service. The Texas Natural Resource Conservation Commission uses yet another watershed layout to determine the availability of water supply, where watersheds are delineated to outlet locations on streams and rivers where permits have been issued for water withdrawal. All of these watershed delineations serve legitimate but different purposes. There are an infinite number of ways to subdivide the landscape into watersheds, and no unique way serves all purposes.

What all these watershed layouts have in common is that they share the same hydro network. In other words, it does not matter how the landscape is subdivided into watersheds once the water reaches the stream and river system. Arc Hydro is designed to allow any set of watersheds

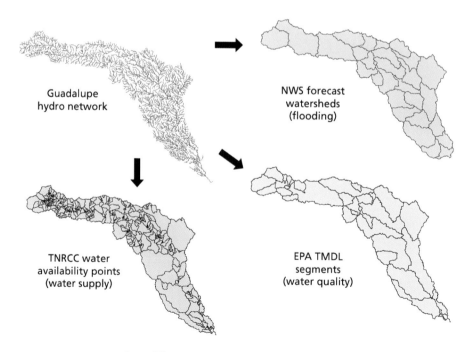

Guadalupe
hydro network

NWS forecast
watersheds
(flooding)

TNRCC water
availability points
(water supply)

EPA TMDL
segments
(water quality)

Four different views of the Guadalupe basin

to be relationally connected to the hydro network using the "area flows to a point on a line" concept to establish relationships between watersheds and hydro junctions at their outlet locations. This approach works regardless of whether the watersheds are derived from digital elevation models or from manual delineation. The flexibility of being able to combine watershed data sets derived from one map source with stream and river networks derived independently is a strength of Arc Hydro. This is a significant advance over previous approaches to working with watersheds and stream networks in GIS that have been almost entirely based on raster analysis of digital elevation models. Indeed, some countries in the world do not have publicly available digital elevation models and their GIS data sets are entirely based on vector watershed and stream networks. Such vector data can readily be incorporated into the Arc Hydro data model.

Basins
Although there is no unique way of defining a watershed layout, many agencies have a standardized set of watersheds that have been worked out over the years, and that serve as reference units for water resources management. In Arc Hydro, these standardized watersheds are called Basins, and they serve as reference units for data management and data packaging. In particular, in applying Arc Hydro, it may be convenient to develop an Arc Hydro geodatabase for each basin. Later in this chapter a method of regionalizing the HydroID is presented. This enables geodatabases defined for a set of basins to be merged to form a geodatabase for a water resource region.

Information sources

Hydrologic units of the United States
In the United States, a standard set of basins called hydrologic units has been developed by the U.S. Geological Survey and indexed by a Hydrologic Unit Code. Hydrologic units are arranged in a hierarchy. At the highest level, the United States is divided into 20 two-digit water resource regions.

Within the two-digit water resource regions are defined four-digit subregions, six-digit basins, and eight-digit subbasins. The eight-digit Hydrologic Unit Code subbasins, popularly known as the "HUCs" or "cataloging units," have become the standard geospatial units for packaging GIS data for water resources in the United States. The eight-digit cataloging units have an average area of 3,700 square kilometers, and there are 2,156 of them within the continental United States, about the same number as the number of counties. Thus the cataloging units can be thought of as "hydrologic counties." They serve some of the same purposes that counties do, namely for archiving and indexing spatially distributed data across the country.

The Guadalupe basin is comprised of four eight-digit cataloging units: the Upper Guadalupe (12100201), Middle Guadalupe (12100202), San Marcos (12100203), and Lower Guadalupe (12100204). The illustrations presented in this chapter focus on the San Marcos subbasin, which is subbasin 12100203 within basin 121002 within subregion 1210 within region 12 in the Hydrologic Unit Cataloging system.

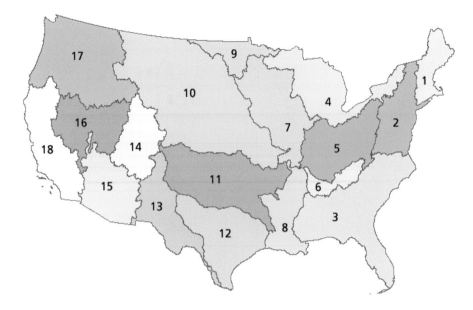

Water resource regions of the contiguous United States

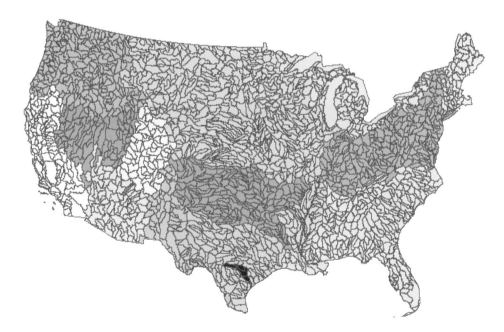

Eight-digit hydrologic cataloging units. The four cataloging units making up the Guadalupe basin are highlighted in red. The colored backgrounds delineate the water resource regions.

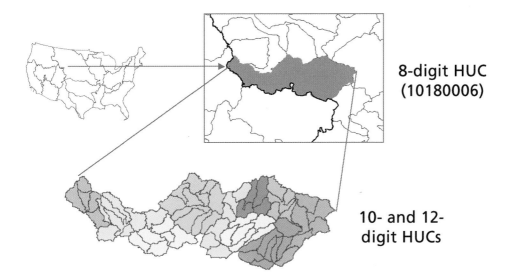

The Watershed Boundary Dataset for Wyoming subdivides eight-digit HUCs into 10-digit HUCs (colored) and then into 12-digit HUCs (outlined).

For water resources management, an area of 3,700 square kilometers is considered large, so there is an ongoing effort by the U.S. Department of Agriculture and the U.S. Geological Survey to create a Watershed Boundary Dataset that further subdivides the eight-digit cataloging units into 10-digit and 12-digit units, called watersheds and subwatersheds, respectively.

Elevation derivatives for national applications
The standard digital elevation model of the United States is called the National Elevation Dataset (NED). It provides a seamless coverage of the nation using one arc-second cells, corresponding on the land surface to a cell size of approximately 30 meters. The illustrations derived from digital elevation models presented in this chapter were developed using this data set. The EROS Data Center, located in Sioux Falls, South Dakota, is presently undertaking a project in collaboration with the National Severe Storms Laboratory in Norman, Oklahoma, to process the National Elevation Dataset into a fine-scale set of catchments. This data set is called the Elevation Derivatives for National Applications, popularly known as EDNA. The name EDNA was chosen in part to emphasize the relationship between terrain information and catchments derived from it, or between NED and EDNA. A five-thousand-cell threshold drainage area (4.5 square kilometers) is used to define the beginning of a stream segment, and catchments are delineated for each stream segment. This threshold drainage area was chosen so that the resulting catchments would be roughly the same size as the standard spatial units for the National Weather Service's Nexrad radar rainfall. In this way Nexrad radar rainfall could readily be mapped onto catchments for flood warnings. The Arc Hydro time series data storage and capacity for downstream

National Elevation Dataset of the United States

accumulation of catchment properties may facilitate the processing of storm rainfall information for this purpose.

The EDNA catchments are classified into a hierarchy of larger to smaller catchment units using the Pfaffstetter classification system. In this system, the highest level drainage areas are subdivided using a set of rules into 10 subdrainage areas, each of which is similarly subdivided into 10 further subdrainage areas, and so on (Verdin and Verdin, 1999).

Watersheds and stream networks in cities

The techniques and data sets described up to this point apply to regional or national information. Yet, the same concepts can be used to apply Arc Hydro within a city or other urban jurisdiction. The Arc Hydro framework data set is simple enough that it can be applied directly to existing digitized streams, watershed boundaries, water bodies, and monitoring points. Achieving a greater degree of detail requires building Arc Hydro data using aerial photogrammetry to identify vector features such as buildings, roads, and streams. Defining stream networks in cities is complicated by the fact that water flows along curbs and drainage ditches that empty into underground storm sewers and then discharge into streams; not all the segments of the urban hydro network are visible on the land surface. Another complication arises where water flow paths intersect with major highway interchanges, and a significant effort is required to define what happens under the highways.

As an aid to its water-quality management program, the Drainage Utility of Austin, Texas, has developed a complete, digitized stream network for the city and all areas draining through the city at a scale of 1 inch to 100 feet, based on interpretation of aerial photogrammetry. This network is coupled with drainage areas derived from the National Elevation Dataset for regional analysis of water quality over the whole city. LIDAR mapping is used to define the terrain surface more precisely for flood damage reduction projects.

International watershed and stream network data

The EROS Data Center has compiled data sets similar to NED and EDNA for the whole earth. The digital elevation model is called GTOPO30, and represents the terrain surface of the earth in 30 arc-second cells (approximately 1 kilometer on the ground surface). The corresponding drainage area and stream network data set is called Hydro1K Derivatives, and like EDNA, it arranges drainage areas in a hierarchy according to the Pfaffstetter system. In addition to catchment and basin boundaries, the Hydro1K data set provides a set of raster data products derived from GTOPO30 to facilitate reprocessing the data to derive watershed boundaries at chosen points in any region of the earth.

The ability provided by GTOPO30 and Hydro1K to delineate drainage areas for any location on the earth is remarkable, and supports the concept that hydrologic models can be applied using GIS data consistently in any country.

Arc Hydro geodatabase for the city of Austin. Hydro response units are based on the land-use map.

Pfaffstetter Level 1 Basins of North America from the Hydro1K data set

Drainage analysis using digital elevation models

Automated delineation of drainage areas is carried out using a model of the land-surface terrain. This model can be a raster digital elevation model (DEM) or a triangulated irregular network (TIN). TIN data consists of irregularly spaced elevation points in x,y,z values derived from land surveying or aerial mapping that precisely represent the terrain surface. Such data is necessary for floodplain mapping, as described in chapter 5, but currently it is cumbersome to edit TIN data for large regions, so that water will everywhere flow downhill over the TIN surface. DEMs more approximately represent the land surface as compared to TINs, and processing DEMs to properly define drainage flow paths is much simpler because of its regular cell structure. Hence DEMs are the most widely used terrain model for drainage delineation. DEMs are digital records of terrain elevations for ground positions at regularly spaced horizontal intervals. This gridded data is derived from the contour information on standard topographic quadrangle maps, or interpolated from irregularly spaced x,y,z points or contours derived from aerial mapping.

All ArcGIS raster operations involved in watershed delineation are derived from the premise that water flows downhill, and in so doing will follow the path of steepest descent. In a DEM grid structure, there exist at most eight cells adjacent to each individual grid cell (cells on the grid boundary are not bounded by cells on all sides). Accordingly, water in a given cell can

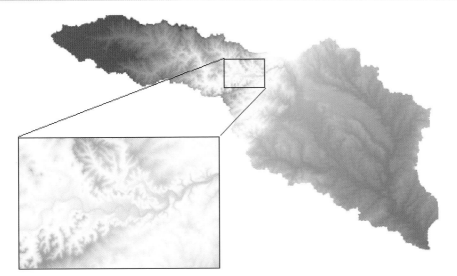

Digital elevation model of the San Marcos subbasin from the National Elevation Dataset

flow to one or more of its eight adjacent cells according to the slopes of the drainage paths. This concept is called the eight-direction pour point model. There are several variants of the eight-direction pour point model, but the simplest, and the one used in ArcGIS, allows water from a given cell to flow into only one adjacent cell, along the direction of steepest descent. The resulting flow direction is encoded 1 for east, 2 for southeast, 4 for south, and so on, to 128 for northeast, a numbering scheme that is derived from the series, which is written in the binary representation used by computers as 00000001, 00000010, 00000100, and so on, to 10000000.

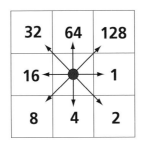

Flow directions in the eight-direction pour point model

Watershed delineation with the eight-direction pour point model is best explained with an illustration. For demonstration purposes, assume a section of a sample DEM grid is provided as shown. The numbers in each grid cell represent the cell elevation.

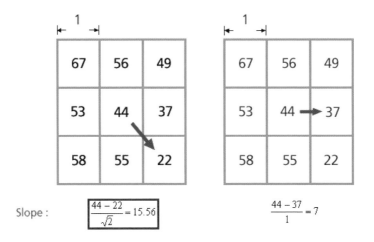

Slope calculations with the eight-direction pour point model. On the left-hand grid, slope is calculated for diagonal cells. On the right, slope is calculated for cells with common sides.

Flow direction grid

The first important grid derived from the digital elevation model grid is the flow direction grid. In the center cell of the illustration (elevation = 44), only two of the eight adjacent cells contain elevations less than 44. This limits the possible flow directions since water will not flow to a cell at a higher elevation. Water will flow in the direction of steepest descent, where slope is defined by elevation decrease per unit travel distance. There are two cases: (1) along the diagonal, where the slope is calculated by subtracting the destination cell elevation from the origin cell elevation, and dividing the result by 1.41 times the cell size, and (2), which applies to water flow in the rectilinear directions through the sides of the cell, where the slope is calculated simply as the elevation difference divided by the cell size. In the case illustrated, the slope along the diagonal is greatest, and water flows to the southeast, so the center cell is assigned a flow direction value of 2. This process is repeated for each of the cells in the DEM grid, thereby creating the flow direction grid whose cell values are the flow directions defined by the eight-direction pour point model. In flat areas, where all surrounding cells have the same elevation as the cell being processed, the search width is expanded until a direction of steepest descent is found. It is important to note that the DEM must have enough precision of elevation measurement to support correct flow direction determination. Large extents of flat areas might produce unnatural drainage patterns. In those cases, a better resolution DEM needs to be used to get satisfactory results.

The flow direction grid can also be graphically symbolized by arrows drawn over each cell, or by a flow network drawn between the cell centers. Even though this grid network is not formally defined by a set of lines in Arc Hydro, raster DEM processing does follow an implied network concept.

67	56	49	46	50
53	44	37	38	48
58	55	22	31	24
61	47	21	16	19
53	34	12	11	12

2	2	4	4	8
1	2	4	8	4
128	1	2	4	8
2	1	4	4	4
1	1	1	2	16

Grid operations. On the left is a DEM grid. The right is a flow direction grid.
The areas in red are from the grids on the previous page.

It is necessary to consider the possibility that flow might accumulate in a cell or set of cells in the interior of the grid, and that the resulting flow network may not necessarily extend to the edge of the grid. An example of such a situation is in the Great Salt Lake in Utah, an inland lake with no outlet to the ocean. A second potential problem arises, moreover, where the DEM grid itself contains artificial pits in the terrain, due to errors in elevation determination or grid development. These artificial sinks must be eliminated in order to accurately delineate watersheds. A pit is where a set of one or more cells is surrounded on all sides by cells of higher elevation.

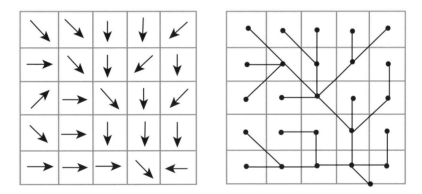

Physical representation of flow direction grids. The left grid has directional arrows,
and the right shows a flow network.

Pits in the DEM are removed through the use of a sink-filling function in ArcGIS. This raises the elevation of all the cells in a pit to the minimum elevation of the surrounding cells so that water can flow across the terrain surface.

Flow direction

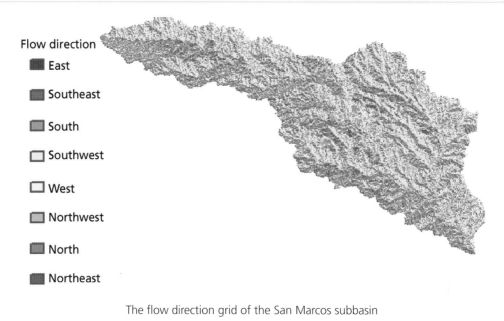

■ East

■ Southeast

□ South

□ Southwest

□ West

□ Northwest

■ North

■ Northeast

The flow direction grid of the San Marcos subbasin

Flow accumulation grid

Flow accumulation is calculated from the flow direction grid. The flow accumulation grid records the number of cells that drain into an individual cell in the grid. Note that the individual cell itself is not counted in this process. From the physical point of view, the flow accumulation grid is the drainage area measured in units of grid cells. The flow accumulation grid for the San Marcos basin clearly shows how drainage area accumulates above the principal rivers and streams of the basin.

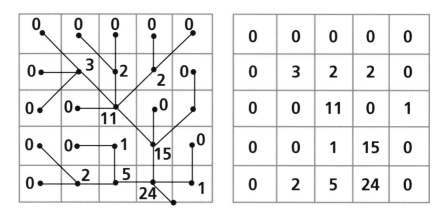

Flow accumulation: number of cells draining into a given cell along the flow network

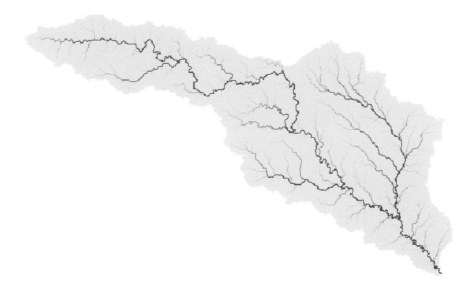

Flow accumulation grid of the San Marcos subbasin.

Stream definition using a threshold drainage area

With a flow accumulation grid, streams may be defined through the use of a threshold drainage area or flow accumulation value. A typical value to use with the National Elevation Dataset is 5,000 cells, which means that all cells whose flow accumulation is greater than 5,000 cells are classified as stream cells, while the remaining cells are considered the land surface draining to the streams. The cell values are assigned 1 where there is a stream and NODATA elsewhere. NODATA is a standard ArcGIS raster notation for a cell with an undefined value. Of course, any cell threshold value may be used, but below a threshold of 1,000 cells the resulting catchment area delineation becomes more questionable, especially in regions of flat terrain. A threshold flow accumulation of 5,000 cells with a 30-meter cell size means that it takes a drainage area of $5,000 \times 30 \times 30 = 4,500,000$ m^2 or 4.5 km^2 to generate a stream.

At this point all the stream cells are labeled identically with a value of 1. The next step is to divide the stream network into distinct stream segments or links. In other words, instead of all the stream cells being 1,1,1,1, etc., for the first link they are labeled 1,1,1, etc., for the second link, 2,2,2, etc., and so on for succeeding links. Links are defined between stream confluences.

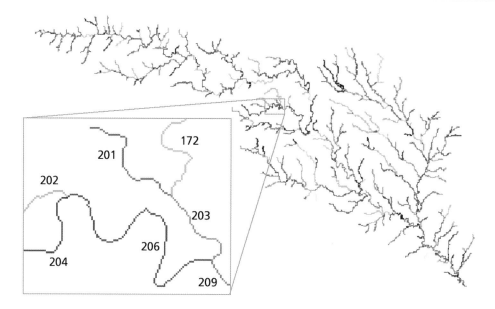

Stream links grid for the San Marcos subbasin. Each link has a unique identifying number.

Catchments

To define catchments for each stream link, we use the flow direction grid to define the zone of cells whose drainage flows through each stream link. The results of the delineation are stored in a catchment grid, whose values are 1,1,1, and so on for cells flowing through the stream link 1 then 2,2,2, and so on for cells flowing through the stream link 2, continuing in this pattern for all links. In the Arc Hydro terrain processing functions, this number is carried forward through the grid operations and is called the GridID. The GridID serves to relate the catchments with the stream links from which they were created. This is the basis of the simple "area flows to line" concept of linking the land surface to the water flow system.

Once the catchment grid is defined, it can be converted into a set of catchment polygons using standard ArcGIS raster-to-vector conversion functions. This process may generate "spurious polygons," which are isolated single cells or small groups of cells connected along a diagonal flow direction with the rest of the catchment. The Arc Hydro toolset contains an automated procedure to detect the existence of such polygons and eliminate them by "dissolving" them into the correct parent catchment polygon. The stream links are also vectorized to form DrainageLines, and the outlet cells are vectorized to form DrainagePoints. The Catchment, DrainagePoint, and DrainageLine feature classes in Arc Hydro store the vector products resulting from terrain analysis using a digital elevation model.

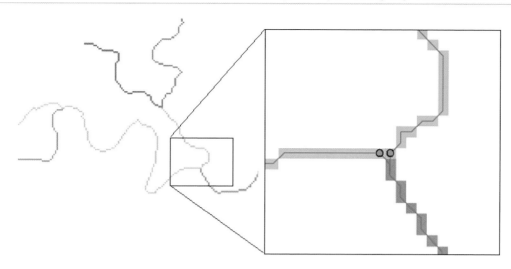

DrainageLines (blue) are drawn through the centers of cells on the stream links. DrainagePoints (magenta dots) are located at the centers of the outlet cells of the catchments.

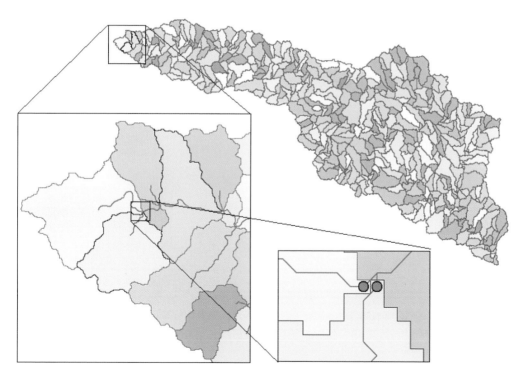

Catchments, DrainageLines, and DrainagePoints of the San Marcos basin

Watershed delineation

Catchment processing from a digital elevation model is a preliminary or preprocessing step to watershed delineation using the Arc Hydro watershed processing tools. The standard application for watershed delineation is to identify a series of points on the hydro network as watershed outlets, then divide and merge the underlying catchments to produce a watershed layout. There are two alternatives: to produce watersheds or to produce subwatersheds. When producing watersheds with the Arc Hydro toolset, the result is a watershed polygon for each outlet point that covers its entire upstream drainage area. Producing subwatersheds results in a polygon for each outlet point that covers only the incremental drainage area upstream of this outlet point and downstream of all others. The subwatershed polygon set is what most people think of when they see a watershed map, and this is what is contained in the Arc Hydro Watershed feature class.

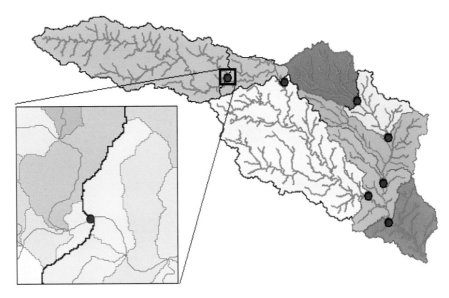

Watersheds delineated from each USGS stream-gaging site within the San Marcos subbasin.
Watershed outlet points may lie within the interior of a catchment. The most downstream watershed is the portion of the basin below the most downstream gaging station.

For any set of drainage areas, there exists three maps that can be produced: the total drainage area upstream of each outlet point, the subdrainage area upstream of this point and downstream of all other outlet points, and what is called in Arc Hydro the adjoint drainage area, which means the difference between the total drainage area upstream and the local subdrainage area. During Arc Hydro processing of catchments, the adjoint catchment is produced for each catchment, to simplify the later processing of watersheds.

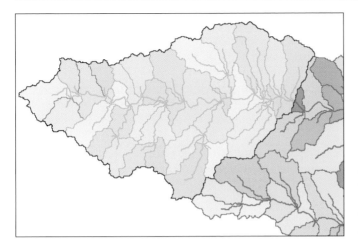

Adjoint catchment: the catchment colored dark green has an adjoint catchment shaded in light green. Together, these features define the total area draining to the catchment outlet.

Watersheds draining to water bodies

The conventional method of delineating catchments and watersheds is to define their outlets by points on a stream network and delineate the area upstream of that point. Similarly, there may be a need to delineate the catchment or watershed of a lake or of a coastal bay segment. These tasks can be accomplished with ArcGIS, but they require special processing. For a lake, shoreline catchments can be delineated for each HydroEdge shoreline segment around the lake. If necessary, these shoreline catchments can be merged with the lake itself to form a single water body catchment.

To generate the total water body watershed, the shoreline catchments are merged with the catchments defined for all the HydroEdge flowlines draining into the water body.

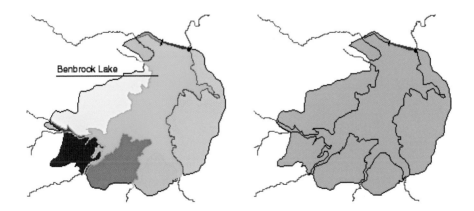

Drainage to a water body. The left map shows catchments of each shoreline segment. The right shows a water body catchment formed by merging the shoreline catchments with the water body.

Watershed draining to a bay segment formed by merging the shoreline catchments with catchments draining tributary streams

Store area outlets

One of the key challenges in assembling disparate data sets in an Arc Hydro geodatabase is that there may be one or several watershed or catchment layouts that the hydrologist wants to connect to the hydro network. These drainage areas may have been developed from a digital elevation model using the Arc Hydro tools, or they may have come from independent sources. In particular, it is a challenge to be able to combine raster-based catchment data sets with vector-based stream networks produced as part of standard hydrographic data sets. Each hydrographic reach may be associated with one catchment or sections of several catchments. Each catchment may have one, many, or even no reaches within it. The Arc Hydro toolset has a Store Area Outlets tool that facilitates linking drainage areas with the hydro network, and contains special provisions for dealing with the complications of combining raster and vector data.

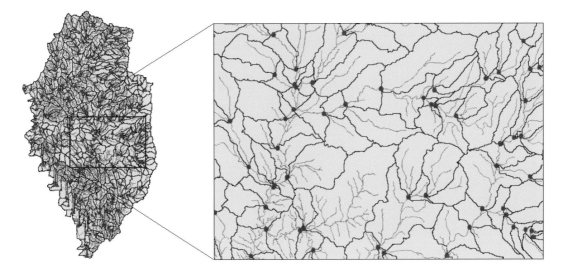

These maps were created using the Arc Hydro Store Area Outlets tool to connect 646 DEM-based catchments with an independently derived vector-based stream hydrography through the use of HydroJunctions at catchment outlets.

Regional analysis

As the precision of digital elevation models increases, their cell size decreases and the number of cells needed to cover a study region increases. The size of the digital elevation model may become so large that it is cumbersome or even impossible to process as a single data set. One solution to this dilemma is to resample the DEM so that it has larger cells, but this reduces the precision of the watershed delineation.

The San Marcos basin requires a grid of 2,500 x 4,300 or about 11 million cells to cover it with the 30-meter cell size of the National Elevation Dataset. This is a reasonable number of cells for raster processing functions to operate efficiently. Typically, a digital elevation model covering one eight-digit hydrologic cataloging unit is reasonable (the San Marcos basin is one such unit). For the Guadalupe basin as a whole (four cataloging units), the required grid size is 70 million cells, which is still workable with a single grid but more cumbersome and time consuming, particularly if an analysis has to be repeated several times because of local changes to the data in one small area of the basin.

The Trinity basin in Texas requires a grid of 237 million cells to be covered, and that size is beyond the realm of reasonable processing times. In fact, completing the data development for a water availability modeling study of the Trinity basin using a single terrain grid required 11 continuous days of processing, and it was not clear if somewhere in the middle of this activity the processing had stopped for some unknown reason. Then there were a few changes to the input data and the whole exercise had to be repeated. This is completely impractical!

The solution to this problem is to combine the Arc Hydro vector and raster processing tools into a regional analysis where the study area is subdivided into subbasins, each of which is

79

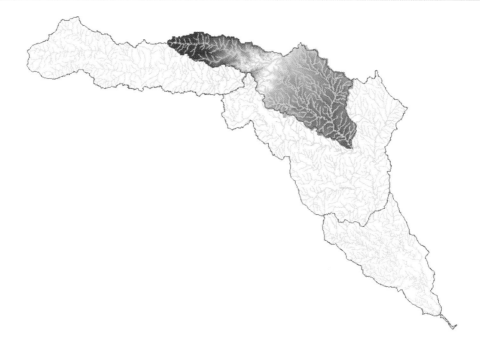

To do regional analysis, the terrain for each subbasin is analyzed separately, and the results are combined by linking them to the hydro network for the whole region.

processed separately with its own smaller digital elevation model. The results are combined to create a regional database.

The process of merging subbasin Arc Hydro geodatabases into a regional Arc Hydro geodatabase is facilitated by special treatment of the HydroID assignment to the features in each subbasin. If, in addition to the feature class number and the object number, the HydroID includes a subbasin number, all features within a subbasin will be uniquely identified within the subbasin geodatabase, and also within the merged regional geodatabase. Relationships between HydroJunctions and Watersheds, Waterbodies and MonitoringPoints formed at the subbasin level will still be valid at the regional level, since the HydroID values used to populate them stay the same.

Combining a regional hydro network with locally delineated drainage areas is a powerful device for regionalization. For determining the upstream drainage area of a particular location within a subbasin, a trace is run upstream on the regional database using the hydro network to select all the upstream catchments for all subbasins, not just the ones in the local subbasin. The properties of the selected catchments can be accumulated downstream across hydro networks so that valid results are obtained both locally within the subbasin and also regionally. For the Trinity basin case, a total of 12 subbasins was used, one for each of the eight-digit hydrologic cataloging units in the basin. The results for the combined network–raster analysis were just as accurate as those obtained by using the laborious method of applying a single digital elevation model grid to the whole Trinity basin.

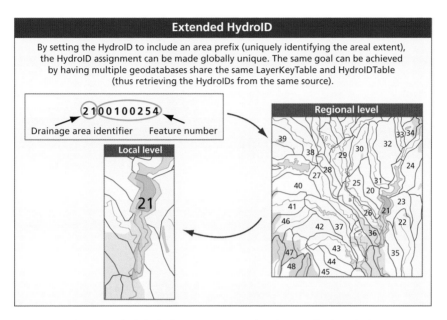

An extended HydroID to support regional watershed analysis

Watershed analysis

Once an Arc Hydro geodatabase has been developed, it may be used to analyze watershed and stream network properties and perform a limited number of hydrologic analyses. A key task for hydrologists is to summarize the properties of catchments or watersheds. This involves in some cases considering the properties of all the drainage areas upstream. Likewise, the outflow from any catchment or watershed affects all the downstream streams and rivers. An elegant capability has been built into Arc Hydro to do area-to-area navigation across drainage areas without having to use the hydro network tracing capabilities. This capability requires that each drainage area possess a NextDownID attribute, which is the HydroID of the next downstream drainage area. Then, using NextDownID on the drainage area, a trace can be run upstream or downstream for a selected drainage area, or the two traces can be combined to identify the "region of hydrologic influence" of a given drainage area, namely the set of upstream and downstream areas that either influence this area or are influenced by it.

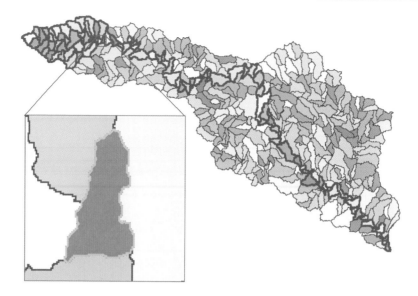

The region of hydrologic influence of a selected catchment in the San Marcos basin

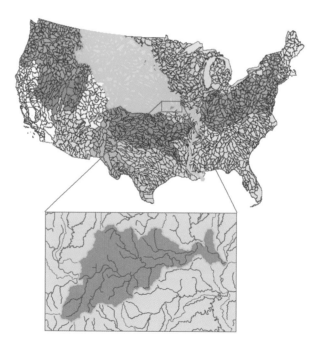

Area-to-area tracing using Arc Hydro USA identifies a region of hydrologic influence for a selected hydrologic unit code watershed in the Missouri basin.

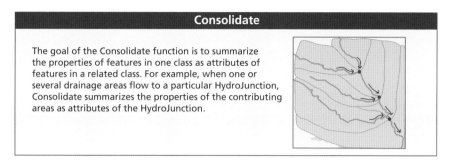

Consolidate

The goal of the Consolidate function is to summarize the properties of features in one class as attributes of features in a related class. For example, when one or several drainage areas flow to a particular HydroJunction, Consolidate summarizes the properties of the contributing areas as attributes of the HydroJunction.

Consolidate

Suppose a given drainage area contains a number of components, such as Nexrad radar rainfall cells or land-use polygons, and we want to summarize their properties and attach the result to the drainage area as a new attribute. For example, we may want to find the total area of each particular land use within the drainage area. The Arc Hydro Consolidate function carries out this task. It assumes that the features to be summarized are connected by a formal relationship between them, for example by using the HydroID of the drainage area as the DrainID of all features in that area.

The Consolidate function uses the operators Average, Sum, Maximum, Minimum, Median, Count, Mode, and Standard Deviation to allow statistics of the selected features to be summarized if necessary. Area-weighted averages of the features can also be consolidated.

Another way to use the Consolidate function is to summarize the properties of all the drainage areas related to a HydroJunction as attributes of that HydroJunction. These might include the local drainage area, runoff, or pollutant loads from that area.

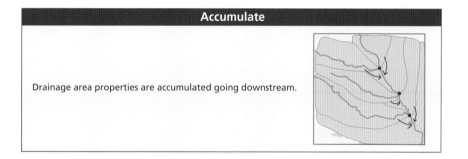

Accumulate

Drainage area properties are accumulated going downstream.

Accumulate

One of the most useful functions in the raster hydrology tools for GIS is the FlowAccumulation function. It sums up, going downstream, the number of cells upstream of each cell, or applies a weight, in which case the weights are summed for all upstream cells. For example, if mean annual precipitation is defined on each cell, it can be treated as a weight and summed using FlowAccumulation, so that for each downstream cell, the total volume of annual precipitation falling upstream of that cell can be determined. In this way the flow to be expected in the stream can be estimated.

In Arc Hydro, the catchments play the same role in the vector domain as cells do in the raster domain. Each catchment has one and only one downstream catchment. Hence, catchment properties can be accumulated downstream using the Arc Hydro Accumulate function, in the same manner as the FlowAccumulation function operated on a raster grid. Indeed, properties can be accumulated going downstream on any feature class for which a NextDownID is defined. This is how the DrainArea attribute of HydroJunctions is populated: the areas of all catchments or watersheds attached to upstream junctions are consolidated onto the junctions, then accumulated going downstream through the junctions to find total drainage area. This function is very useful for doing steady-state hydrologic analysis, in particular for defining maps of mean annual runoff and pollutant loadings for rivers and water bodies, which are needed for Total Maximum Daily Load analysis of water quality.

For more complex watershed analyses involving attributes varying through time, the Arc Hydro time series component and modeling methods using Visual Basic should be used. For more details, see chapters 7 and 8.

Data dictionary

These diagrams summarize the object and feature classes in the Drainage component of Arc Hydro, and their interrelationships. All the classes shown are available for loading data because they have inherited all the attributes from classes located above them in the UML hierarchy. The attributes shaded in blue are ESRI standard attributes, while those shaded in white are Arc Hydro attributes. The terms used here are defined in the glossary at the back of this book.

Simple feature class **Basin**					Geometry *Polygon* Contains M values *No* Contains Z values *No*				Basins are a set of administratively selected standard drainage areas usually named after the principal streams and rivers of a region.

Field name	Data type	Allow nulls	Default value	Domain	Prec- ision	Scale	Length	
OBJECTID	OID							
Shape	Geometry	Yes						
HydroID	Integer	Yes			0			Unique feature identifier in the geodatabase
HydroCode	String	Yes					30	Permanent public identifier of the feature
DrainID	Integer	Yes			0			HydroID of the reference drainage area feature
AreaSqKm	Double	Yes			0	0		Area in square kilometers
JunctionID	Integer	Yes			0			HydroID of the HydroJunction at drainage outlet
NextDownID	Integer	Yes			0			HydroID of the next downstream basin
Shape_Length	Double	Yes			0	0		
Shape_Area	Double	Yes			0	0		

See network features

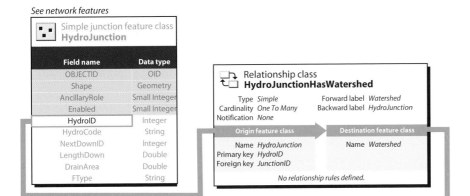

Simple junction feature class
HydroJunction

Field name	Data type
OBJECTID	OID
Shape	Geometry
AncillaryRole	Small Integer
Enabled	Small Integer
HydroID	Integer
HydroCode	String
NextDownID	Integer
LengthDown	Double
DrainArea	Double
FType	String

Relationship class
HydroJunctionHasWatershed

Type	*Simple*	Forward label	*Watershed*
Cardinality	*One To Many*	Backward label	*HydroJunction*
Notification	*None*		

Origin feature class		Destination feature class	
Name	*HydroJunction*	Name	*Watershed*
Primary key	*HydroID*		
Foreign key	*JunctionID*		

No relationship rules defined.

Simple feature class
Watershed

Geometry	*Polygon*
Contains M values	*No*
Contains Z values	*No*

Field name	Data type	Allow nulls	Default value	Domain	Precision	Scale	Length
OBJECTID	OID						
Shape	Geometry	Yes					
HydroID	Integer	Yes			0		
HydroCode	String	Yes					30
DrainID	Integer	Yes			0		
AreaSqKm	Double	Yes			0	0	
JunctionID	Integer	Yes			0		
NextDownID	Integer	Yes			0		
Shape_Length	Double	Yes			0	0	
Shape_Area	Double	Yes			0	0	

Watersheds are drainage areas defined by subdividing the landscape into units convenient for a particular analysis.

Unique feature identifier in the geodatabase
Permanent public identifier of the feature
HydroID of the reference drainage area feature
Area in square kilometers
HydroID of the HydroJunction at drainage outlet
HydroID of the next downstream watershed

Simple feature class
Catchment

Geometry	*Polygon*
Contains M values	*No*
Contains Z values	*Yes*

Field name	Data type	Allow nulls	Default value	Domain	Precision	Scale	Length
OBJECTID	OID						
Shape	Geometry	Yes					
HydroID	Integer	Yes			0		
HydroCode	String	Yes					30
DrainID	Integer	Yes			0		
AreaSqKm	Double	Yes			0	0	
JunctionID	Integer	Yes			0		
NextDownID	Integer	Yes			0		
Shape_Length	Double	Yes			0	0	
Shape_Area	Double	Yes			0	0	

Catchments are elementary drainage areas defined by subdividing the landscape according to a set of physical rules.

Unique feature identifier in the geodatabase
Permanent public identifier of the feature
HydroID of the reference drainage area feature
Area in square kilometers
HydroID of the HydroJunction at drainage outlet
HydroID of the next downstream catchment

Simple feature class **DrainagePoint**				Geometry	*Point*			
				Contains M values	*No*			
				Contains Z values	*No*			
Field name	Data type	Allow nulls	Default value	Domain		Prec-ision	Scale	Length
OBJECTID	OID							
Shape	Geometry	Yes						
HydroID	Integer	Yes				0		
HydroCode	String	Yes						30
DrainID	Integer	Yes				0		
JunctionID	Integer	Yes				0		

A point at the center of a DEM cell which is the outlet of a DEM-derived drainage area.

Unique feature identifier in the geodatabase
Permanent public identifier of the feature
HydroID of the reference drainage area feature
HydroID of the HydroJunction at drainage outlet

Simple feature class **DrainageLine**				Geometry	*Polyline*			
				Contains M values	*No*			
				Contains Z values	*No*			
Field name	Data type	Allow nulls	Default value	Domain		Prec-ision	Scale	Length
OBJECTID	OID							
Shape	Geometry	Yes						
HydroID	Integer	Yes				0		
HydroCode	String	Yes						30
DrainID	Integer	Yes				0		
Shape_Length	Double	Yes				0	0	

A line drawn through the center of cells on a DEM-derived drainage path.

Unique feature identifier in the geodatabase
Permanent public identifier of the feature
HydroID of the reference drainage area feature

River channels

Nawajish Noman, ESRI
James Nelson, Brigham Young University

Because of the importance of rivers to human activity and quality of life, engineers and scientists pursue an increased understanding of rivers' physical properties and behaviors. A wealth of data has been collected about rivers, including their morphology, water quantity and quality, surrounding habitats, and flooding potential. The Arc Hydro data model provides a way to efficiently store, manage, and retrieve this vital river system data. Arc Hydro provides a standardized way to represent channel profiles and cross sections, making it easier to visualize and extract channel information in ArcGIS.

Understanding rivers

Rivers have been a focus of human activity throughout ancient and modern times. So important to humanity are the benefits obtained from rivers, and so necessary is the protection against floods and other river disasters, that the study of riverine systems has advanced in leaps and bounds. Engineers and scientists have been fascinated by the self-formed geometric shapes of rivers, as well as their complex ecosystems and responses to natural changes and human interference. It is clear that rivers, as a part of nature, can be mastered not by force but by understanding.

Rivers are studied to understand water supply, channel design, flood control, navigation improvement, water quality, regulation, and more. Yet no matter what the purpose is, the success of any river study largely depends on the volume of data available and the tools available for data storage, retrieval, and analysis. In recent years GIS has become an excellent tool for spatial data storage, visualization, and analysis.

Producing a cross section of a river channel is fundamental to all river studies. Whether a hydrologist needs to find the discharge or examine the profile of a feature such as a meander or riffle, it is necessary to produce a cross section of the river. Arc Hydro provides a framework for storing channel and cross-section information in a systematic manner. In Arc Hydro the river channel is considered a three-dimensional network of cross-section lines located transverse to the channel and profile lines drawn parallel to it. This simple and effective way of storing and extracting channel and cross-section data makes Arc Hydro suitable for applications such as river morphology, floodplain delineation, and river modeling.

River morphology

One way to track the amount and location of erosion and deposition in a river valley is to establish a series of cross sections through the valley and survey the cross sections regularly, especially before and after periods of heavy rain and flooding.

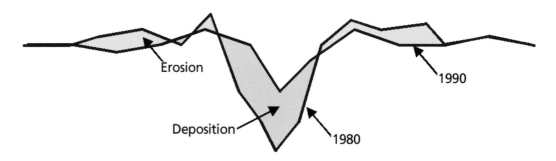

This cross-section comparison determines the areas of erosion and deposition in the North Fork Toutle River.

One of the most dramatic examples of rapid channel incision and widening in the North Fork Toutle River occurred after the eruption of Mount St. Helens. In only two years, a channel 350 meters wide and 40 meters deep was carved into the landslide deposit that filled the river valley to a depth of 195 meters.

Floodplain delineation

Because of its devastating nature, flooding poses serious hazards to human populations in many parts of the world and the economic damages from floods have increased considerably in the last 30 years. The "Flood Disaster Protection Act of 1973" required that all floodplains in the United States be identified and that flood-risk zones within those areas be established. Water resources engineers have developed methods for delineating floodplain boundaries.

A floodplain is the normally dry land area adjoining rivers, streams, lakes, bays, or oceans that is inundated during flood events. Flooding is caused by the overflow of streams and rivers and abnormally high tides resulting from severe storms. The floodplain can include the full width of narrow, steep stream valleys, or broad areas along streams in wide, flat valleys. The channel and floodplain are both integral parts of the natural conveyance of a stream. The floodplain carries flow in excess of the channel capacity. The greater the discharge, the greater the extent of inundation.

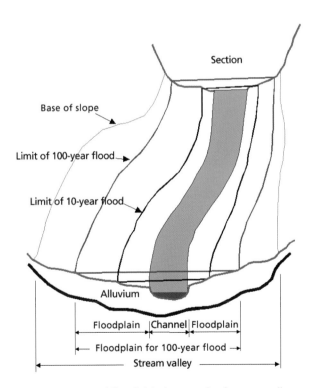

Typical sections and floodplain in a reach of stream valley

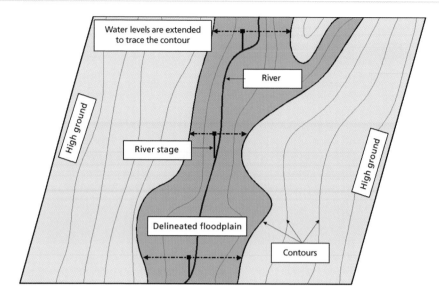

Delineation of floodplains

An automated floodplain delineation process determines inundation extent by comparing simulated water levels from a river hydraulic model with ground-surface elevations. Cross sections are required to represent channel geometry in a river hydraulic model. The accuracy of simulated water levels, and eventually the accuracy of floodplain delineation, largely depends on the shape and the extent of these cross sections. In a flood model, it is important to specify a detailed cross-section geometry that not only extends over the floodplain, but is also capable of carrying the total flood discharge through it.

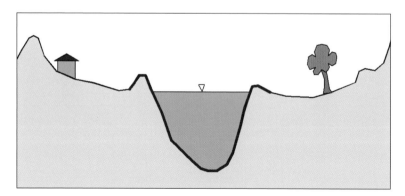

Flow in the main channel

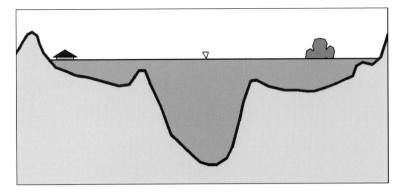

Flow in the main channel and floodplains

River modeling

River hydraulic modeling provides a tool that hydrologists use to study and gain an understanding of hydraulic-flow phenomena, to select and design sound engineering projects, and to predict extreme flooding situations and provide advance warning of their occurrence.

The essential quality of a river hydraulic model is its predictive capacity. For model predictions to be accurate and useful, the model is based on hydraulic equations that represent the most important flow phenomena. But even if the flow equations used are appropriate, the model is unreliable unless correct hydraulic and topographic features are represented in a sound manner. This is perhaps the distinguishing feature of river modeling—the modeler must provide a numerical description of physical reality that is consistent with the physical laws governing water flow, and also consistent with the shape and properties of the river channel. Clearly, the accuracy of these numerical descriptions depends first and foremost on the appropriate schematization of a river network and the quality of the data used in the model.

Data requirements

Theoretically, any physical situation can be simulated in a river model with as high an accuracy as desired within the limits of the validity of the flow equations. The data required for river models can be grouped into two classes: hydraulic and topographic.

Hydraulic data consists of continuous measurement of such things as discharge hydrographs, stage (or water surface elevation), tidal records, spot measurements of stage, continuous discharge and velocity, and rating curves relating stage and discharge.

Topographic data describes the geometry of the simulated river system, meaning the data supplies the elements necessary to define width, cross-sectional areas, and volume of inundated floodplains. Moreover, the data should permit the establishment of the topology of the model: the definition of cells in inundated areas, channel loops, the characteristic cross sections along channels where the computational points are to be established, the limits between main channels and floodplains, and the network of discharge exchange between the cells. The topography of river valleys can be measured with an accuracy and completeness that is limited only by cost, and can be directly carried over into the precise definition of cross sections.

The topographic data used to build river hydraulic models may be divided into two basic categories: qualitative and quantitative.

Qualitative data provides a reconnaissance type of description of the river, its tributaries, and inundated floodplains. Qualitative data identifies the physical conditions that determine flood-development patterns—for example, the existence of berms within the floodplain, dykes, breach information, elevated roads, localized obstacles within inundated zones, and preferential flow axes. Qualitative data can be obtained by field investigation, inquiries, satellite and aerial photographs, and newspaper reports.

Quantitative topographic data is needed for the model representation of the river and its flooded plains. Three essential kinds of quantitative topographic information required in river hydraulic models are: longitudinal profiles along banks, dykes, and roads; cross-section or transverse profiles across the water course; and contour imagery of the inundated area.

It is impossible to completely enumerate all the topographic data needed, since the more that is known, the better. However, the accuracy of model results does not depend on this data alone. If the model does not require or is not capable of evaluating detailed information, there is little benefit in putting that data in the model. Two widely used hydraulic models, HEC-RAS and MIKE 11, developed in different parts of the world, describe the requirement for channel information in river hydraulic models.

HEC-RAS, developed by the U.S. Army Corps of Engineers Hydrologic Engineering Center (HEC), is designed to perform one-dimensional hydraulic calculations for a full network of natural and constructed channels. The current version of the HEC-RAS system supports steady and unsteady flow water-surface profile calculations. The basic computational procedure is based on the solution of the one-dimensional continuity and momentum equations of water flow.

One of the major steps in developing a hydraulic model with HEC-RAS is to enter the necessary geometric data, which consists of connectivity information for the stream system, cross-section data, and hydraulic-structure data. Cross-section data represents the geometric boundary of the stream. Cross sections are required at representative locations throughout the stream, at locations where changes occur in discharge, slope, shape, roughness, and at hydraulic structures. The required information for a cross section includes the river reach and river station identifiers, a description, station and elevation points, and Manning's roughness.

MIKE 11, developed by DHI Water and Environment in Denmark, is a system for the one-dimensional modeling of rivers, channels, and irrigation systems, including rainfall-runoff, advection-dispersion, morphological, water quality, and two-layer flow modules.

The MIKE 11 hydrodynamic module (HD) uses an implicit, finite difference scheme for the computation of unsteady flows in rivers and estuaries. The hydrodynamic module can describe subcritical as well as supercritical flow conditions through a numerical scheme that adapts according to the local flow conditions (in time and space). The formulation can be applied to looped networks and quasi two-dimensional flow simulation on floodplains.

River branches are represented in MIKE 11 models by prescribing the shapes of river cross sections and their locations along the river axes. Floodplains and storage areas are represented in one of two ways: either by including the properties of the floodplain within the specification of each individual cross section, or by representing the floodplains as separate flood cells and routing channels that are connected to river(s) either directly or via hydraulic structures such as weirs or flow regulators. The model allows for two different types of bed-resistance descriptions, Chezy and Manning, and uses this information for flow computation.

Traditionally, channel information is collected in the field using different surveying techniques and tools. First, horizontal and vertical controls are established, then elevations are taken along a cross section. Yet collecting data in this manner is time consuming and costly. Recent development of satellite-based survey techniques and the popularity of GIS software has opened an enormous possibility of extracting cross-section data from digital terrain models (DTMs). DTMs are becoming a major data source for creating cross sections and supplementing surveyed cross sections.

Digital terrain models (DTMs)

A digital terrain model is a surface model representing the topographic surface. In DTMs, ground elevations are stored using formalized data structure rules, either as a triangulated irregular network (TIN) or a grid. Each model has advantages and limitations. A grid is a simpler model of a surface, and digital terrain data is widely accessible in grid format. TINs can produce a more accurate representation of surfaces and features, but usually require a more extensive data collection effort using aerial photogrammetry or other remote-sensing techniques, such as Light Detection and Ranging (LIDAR). This section examines only the TIN representation of a surface.

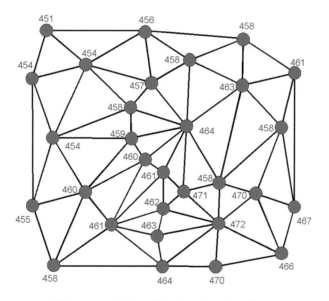

TIN representation of land surface elevation

In recent years, advances in GIS have provided many tools for extracting cross-section data from digital terrain models. Since the TIN surface model allows variable point density based on the degree of change of slope, it actually captures and represents surface features such as streams, ridges, and peaks better than a grid surface model. In TINs these surface features are stored with precise coordinates and provide more accurate vertical profile information. Therefore, a TIN is preferred as a cross-section data source over the grid surface model for representation of riverine topography.

The TIN representation of a surface

TINs represent surfaces as contiguous nonoverlapping triangular faces. A surface value for any location can be estimated by interpolation of elevations within a triangle. Because elevations are irregularly sampled spatially in a TIN, this data structure allows a variable point density in areas where the terrain changes sharply. This yields an efficient and accurate surface model.

In GIS, TINs are made from mass points, which are points with elevations collected from a variety of sources. From these input points, a triangulation is performed. In a TIN, the triangles are called faces, the points become nodes to a face, and the lines of faces are called edges.

A TIN preserves the precise location and shape of surface features. Areal features such as lakes and islands are represented by a closed set of triangle edges. Linear features such as ridges are represented by a connected set of triangle edges. Peaks and pits are represented by triangle nodes. TINs support a variety of surface analyses such as calculating elevation, slope, and aspect, performing volume calculations, and creating profiles on alignments. The disadvantage of TINs compared to grid DEMs is that they are often not readily available and therefore require costly data collection.

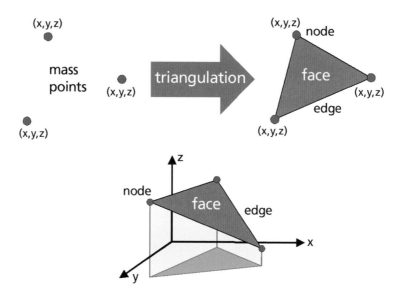

Creating a TIN from mass points

Creating TINs

Given a set of points, many possible triangulations can be created. ArcInfo uses an algorithm called Delaunay triangulation to optimize how faces model a surface. The basic idea of this algorithm is to create triangles that collectively are as close to equilateral shapes as possible. This keeps the interpolation of elevations at new points in closer proximity to the known input points.

TINs can be created by entering surface features that represent elements of terrain such as point elevations, peaks, streams, and ridges, and also man-made features such as roads and embankments. Point elevations are the predominant input into a TIN and form the overall shape of the surface. They can be input from contour lines if necessary, but it is better to use points collected with surveying or photogrammetric devices because the operator can do a better job of visually sampling points that reflect terrain relief. Stream, ridges, embankments, and similar surface features extracted from topographic maps or digital orthophotos are then added to refine the surface model and sharpen the changes in relief. These features are preserved in the TIN and increase the model accuracy.

The features that represent and preserve the surface morphology in a TIN are point surface, linear surface, and areal surface.

TIN features

Mass points represent points at which z-coordinates are measured. After triangulation, they are preserved as nodes with the original location and elevation.

Breaklines are linear features that represent natural features such as streams and ridges, or man-made features such as roadways. Hard breaklines represent a slope discontinuity such as a stream course. While the surface is always continuous, its slope may not be. On the other hand, soft breaklines only add the imprint of a linear feature without representing a slope discontinuity.

Areal surface features are polygons that represent objects such as lakes or coasts. For example, replace polygons assign a single z-coordinate to the boundary and all interior heights, while erase polygons mark all areas within a polygon as being outside the zone of interpolation for the model. Analytical operations such as volume calculations, contouring, and interpolation will ignore these areas.

Data for TINs is commonly compiled with photogrammetric instruments that sample elevations from pairs of aerial photographs precisely aligned in a stereo model. TINs are also produced from survey data, digitized contours, grids with z-coordinates, point sets in files or databases, or operations on other TINs.

The process of extracting cross-section data from a TIN involves defining the TIN as a surface and digitizing cross-section cutlines. A cutline represents the alignment of a cross section in an x,y coordinate or plan view. The ground elevation beneath each vertex is then interpolated from the surface. The elevations are either stored with the cross-section cutlines as 3-D lines or exported to a table.

95

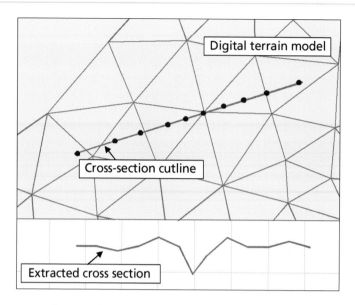

Cross-section extraction from a digital terrain model

Combining the topographic data extracted from digital terrain models with field information results in a better interpretation of land-surface features.

River, channel, and cross section

Even though water resources engineers all over the world deal with similar elements of nature and work to solve similar problems, how they define these elements and address problems depends on engineering practices. Therefore, it is important to define a river or stream channel and a channel cross section.

A river or stream channel is a conduit or water course carrying water flow under gravity. The water surface profile is sloping and the flow has a significant velocity. The width of the channel is much smaller than its length, and the flow is essentially one-dimensional in the direction of the channel centerline. The channel includes the flow in the main channel of the river or stream itself and also in its floodplains to the left and right of the main channel. The channel is like a "cradle" that carries the flow. It has a complex three-dimensional geometry, and additional properties such as channel roughness. Examples of hydrographic features with channels include rivers, streams, creeks, canals, ditches, culverts, and storm sewers.

A cross section as used in this chapter refers to the section of a channel taken transverse to the direction of the flow. Natural channel sections are by nature irregular. For streams subject to frequent floods, the channel may consist of a main channel carrying normal discharges and one or more side channel sections for accommodating overflows.

Three-dimensional view of a channel

The geometry of a cross section is represented by a series of points specified by a pair of m, z coordinates. The m-coordinate denotes the measure, or distance of the point along the cross section from one end. The ground elevation above a datum is denoted by the z-coordinate. The thalweg is defined as the lowest point of the main discharge-carrying portion of a cross section. A cross section is typically identified by the channel name and station (a measure value along the channel flowline) and located by geographic coordinates.

Computations involving flow in open channels commonly require an evaluation of the resistance characteristics of the channel. The Manning and the Chezy equations have been used extensively as an indirect method for computing discharge or depths of flow in natural channels. In 1889 the Irish engineer Robert Manning presented a formula, which was later modified to its present well-known form:

$$V = (1.49/n) \, R^{2/3} S^{1/2}$$

Where V is the mean velocity in feet per second, R is the hydraulic radius in feet (cross-sectional area divided by the length of the wetted perimeter), S is the slope of energy grade line (often taken equal to the bed slope), and n is the coefficient of roughness, known as Manning's n.

Since the Manning and Chezy equations both require bed roughness along with cross-section parameters such as area or hydraulic radius, it is useful to define channel roughness as a parameter of cross sections.

The alignments of a river network together with the cross section's location, geometry, and properties provide the basic channel information for water resources professionals. The following section describes how this information is conceptually represented in GIS using Arc Hydro.

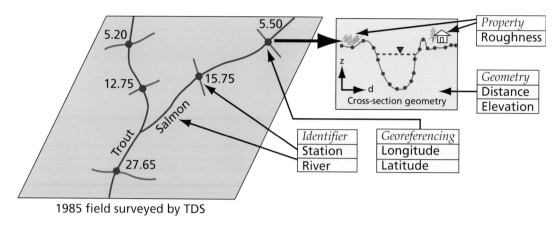

1985 field surveyed by TDS

A channel cross section

Representing channel information with Arc Hydro

The part of Arc Hydro that stores and manages channel information consists of two feature classes and one object class. Feature classes are derived from the ChannelFeature abstract class, which inherits properties from the HydroFeature class. The feature classes are CrossSection and ProfileLine. The object class CrossSectionPoint is derived from ArcGIS object. Collectively, these classes provide a standardized way of representing the channel profiles and cross sections, and make it easier to visualize and extract channel information in ArcGIS.

ChannelFeature
The purpose of the ChannelFeature class is to gather attributes that are common to channel features such as thalweg and cross sections. Its attributes are ReachCode and RiverCode. All channel features derived from this abstract class automatically inherit these attributes.

ReachCode is an identifier that tags each water feature uniquely within the drainage systems of the United States. The NHD reach code is a 14-digit number comprised of the eight-digit Hydrologic Unit Code (HUC) unit containing the reach and a six-digit segment number sequentially assigned within that HUC unit.

RiverCode is another identifier of the river, defined by the river name or by the concatenation of the latitude and longitude of its outlet location. In hydraulic modeling, it is common to use both a ReachCode and a RiverCode to identify a particular segment of a river. For example, the RiverCode could be Walnut Creek, and the ReachCode could be North Branch, Main Stem, or South Branch, to indicate which segment of the Walnut Creek channel is being considered.

Cross section shapes
The shape a cross section can take in a GIS primarily depends on how it is georeferenced. Cross sections digitized as cutlines in a GIS or surveyed as a series of x,y coordinates may form a polyline representing the true alignment of the cross section. Since elevations are measured or

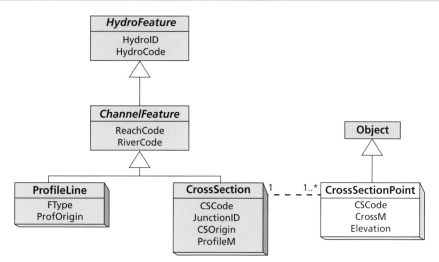

Channel representation in Arc Hydro

extracted from a DTM for each of the points on the cross section, the cross section actually forms a 3-D line where each vertex has x-, y-, and z-coordinates.

As mentioned earlier, DTMs are becoming a major data source for creating cross sections. The process requires obtaining elevation values from a DTM defined as a surface at all vertex locations of the line that represents the alignment of the cross section. This can be achieved in two ways: (1) cross-section cutlines digitized as 2-D lines can be used to extract elevations from a DTM and to output a layer of 3-D lines, or (2) 3-D lines are created as the line is being digitized. In both cases the river network layer and the DTM, along with other geographic features, are displayed in the background to help with digitizing the cross sections at desired locations.

ArcGIS allows an m-coordinate, or linear measure, to be assigned at each point in a point, multipoint, polyline, or polygon. This means that a point may have x-, y-, z-, and m-coordinates. The built-in feature geometry system has functions to interpolate an m-coordinate for the x,y coordinates along a path or to calculate x,y coordinates from an m-coordinate along a path. By taking advantage of this additional geometric attribute, the distance of each point along the cross section can either be assigned or computed and stored as m-values. As a result, 3-D line cross sections with x-, y-, z-, and m-coordinates are readily available for exporting and displaying in the traditional distance-elevation format. It also becomes easier to transform a cross section with distance and elevation values into a 3-D line.

In Arc Hydro, CrossSections are linear features that define the shape of the channel transverse to the direction of river flow. A CrossSection line has vertical measure (z-coordinate) and linear measure (m-coordinate) values at each vertex. Vertical measure (z-coordinate) represents the elevation of channel bottom with respect to a datum such as mean sea level. Linear measure (m-coordinate) represents the distance of each vertex along the cross section. Usually distances are measured from the left end of a cross section, where left is interpreted by looking in the downstream direction of flow in the channel.

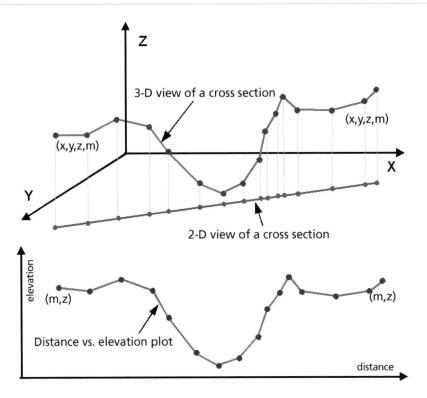

Three-dimensional view of a cross section

The basic attributes of a CrossSection are: CSCode, JunctionID, CSOrigin, and ProfileM.

CSCode is a unique user-defined cross-section identifier.

JunctionID of a cross section identifies the HydroJunction at the equivalent hydrologic location as the cross section even though the precise geographic location may vary slightly.

CSOrigin identifies the source of data and the method used to capture the shape of the cross section. It may also include information on how a surveyed cross section is transformed into a CrossSection polyline feature with linear measure (m) and elevation (z) values.

ProfileM is the station value of a cross section. The value of ProfileM is used to locate the position of the cross section using linear measure (m-coordinate) along a ProfileLine. Depending on the convention, the station values are measured either from the upstream or downstream end of the channel. However, the convention used for ProfileM must be consistent with the direction of ProfileLine measure.

Additional attributes can be added to satisfy the specific requirement of an organization or a particular project. For example, a DatumAdjustment attribute may store correction values that could be used to transfer elevations to a common datum before exporting to a river hydraulic model. A StructureCode will relate a hydraulic structure with the cross section. Special notes regarding a cross section such as ecological significance, accuracy of the measurement, or hydraulic condition of the channel at that location can be stored in a Remark field. These are not included as part of the basic Arc Hydro data model but can easily be added by the user.

CrossSectionPoint

CrossSectionPoint object class stores traditionally surveyed cross sections for which the m- and z-values are known but the x,y locations of the points are not known. A CrossSectionPoint class is relationally connected using CSCode to a CrossSection line feature, which may just be drawn as a marker line across a channel if the station value is known.

The basic attributes of a CrossSectionPoint are CSCode, CrossM, and Elevation.

CSCode is a user-defined cross-section identifier. This identifier is unique for each cross section. This is the "key field" through which a CrossSectionPoint is connected to a CrossSection feature line.

CrossM denotes the location of a CrossSectionPoint along a cross section at which the elevation is known.

Elevation stores the elevation of a CrossSectionPoint above a datum such as mean sea level.

While a single CrossSection line represents a cross section, it requires several CrossSection-Points to represent the geometry of a cross section. Therefore, a one-to-many (1 . . *) relationship exists between a CrossSection feature line and CrossSectionPoints. In order to avoid confusion, the elevation (z-coordinate) and measure (m-coordinate) field values of a CrossSection feature line should be set to NaN (not a number) if the cross-section geometry is stored as CrossSection-Points.

CrossSection events

The geometry of a cross section is not sufficient for most water resources applications. For instance, a river model for hydraulic simulations requires roughness values along a cross section as well as the location of the left bank and the right bank. Cross-section properties such as roughness values, land-use type, zone type, and the locations of the left bank, right bank, thalweg, left floodplain, and right floodplain can be stored in a table corresponding to measure values along a cross section. If necessary, these locations (points) and a part of the cross section (segments) can be displayed in GIS as "point events" and "line events" respectively using dynamic segmentation or linear referencing utilities.

To be useful, all information associated with a cross section, including geometry and properties, must be linked. This is achieved by maintaining a "key field" in the "tables." By establishing a relation using the "key field," one can extract properties information from the property table for desired cross sections.

The HydroEvent objects such as HydroPointEvents or HydroLineEvents are used to define properties at particular points or along particular regions of the cross section. Examples of these events include the location of a left bank, or channel roughness along a section. The CSCode, instead of the ReachCode, should be used as the "primary key" to maintain the relationship between a CrossSection and its events.

The location of a cross-section event alone is not very informative unless we know the type of event and, if applicable, the value of the property. Consider the line event that represents the change in channel-bed roughness along a cross section. We must know roughness values in order to make proper use of this information. However, it is almost impossible to make a

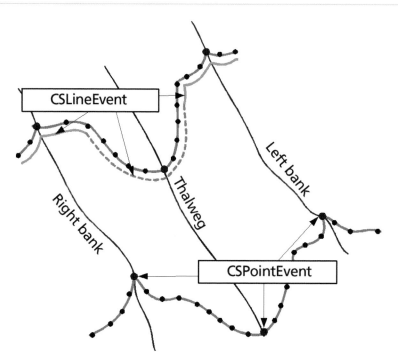

Cross-section properties as point and line events

comprehensive list of the types or properties and associated information, and to make it a part of the basic model. Users can add additional attributes as necessary.

ProfileLines

To determine the flow pattern through a river system, we need to combine the geometries and properties of cross sections with the channels they represent. We also need to relate other features such as embankments and the extent of inundation with the main channel itself and see everything as an integrated system rather than as individual components. A longitudinal profile of a channel provides useful information about breaches and identifies the locations of possible overspilling. Historical flood extent information is very useful for checking the validity of a flood model. In Arc Hydro the longitudinal view of a channel is represented as ProfileLines.

ProfileLines are linear features that define the longitudinal profile of the channel parallel to the direction of flow (e.g., thalweg, left bank, right bank, left floodline, and right floodline). The left floodline and right floodline represent the extent of inundation in the left and right floodplains respectively.

In addition to the properties inherited from the ChannelFeature class, a ProfileLine has two more attributes, FType and ProfOrigin.

FType is the code specifying the feature represented by a ProfileLine, such as thalweg, bankline, or floodline.

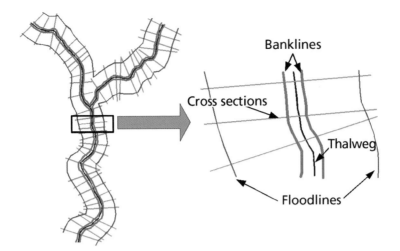

Features of a channel network

ProfOrigin identifies the source of data and the method used to capture the shape of the ProfileLine.

Banklines can be digitized from high-resolution digital orthophotos or maps. Despite the fact that field surveying is expensive and time consuming, it is not uncommon to survey the alignment of the embankments using GPS and other tools. An alternative approach is to display the bank positions as events on cross sections and then digitize banklines by connecting those event points along the channel. A similar approach could be applied for floodlines. Historical records, satellite and radar images, simulation results of extreme events from river hydraulic models, and so on, are valuable sources of inundation extent.

Depending on the scale of the map, it has become customary to represent narrow rivers by a single blue line, whereas wide rivers are drawn as polygons filled in blue and bounded by left

The thalweg line represents the flowline through channels.

103

and right banklines. Water bodies such as lakes are usually represented by polygons. However, in practice, the schematic diagram of a river network is drawn using flowlines. A flowline is a line through the center of a channel reach or a water body that defines the main direction of flow, as discussed in chapter 3. Flow properties such as discharge, water-surface elevation, and constituent concentrations can be attached to locations on the flowline. Since the water by nature flows though the lowest points of the river, it makes sense to use the thalweg line as the flowline. Therefore, the thalweg profilelines in this model also serve the purpose of flowlines. Once the linear measure values (m-coordinates) are assigned from the network, the flow properties (such as discharge and salinity level) can be attached to a thalweg profileline. The thalweg profileline is then used to assign station measures to cross sections.

Deriving channel information

The suitability and success of a data model largely depends on the ease and efficiency of the data-retrieval process. To derive and export channel information, users need to display all the features of a channel along with cross-section geometry and properties, plus draw longitudinal profiles along a channel.

Since ProfileLine and CrossSection are georeferenced feature layers, they are automatically overlayed in a GIS. When multiple representations of the same profile type exist in the database, the attribute ProfOrigin comes into play. For example, if a channel shifts through time, and these alignments are stored in the database, then each alignment has a different ProfOrigin. The user can then display embankments, cross sections, or whatever else may be relevant to a particular channel alignment by selecting the features using ProfOrigin and the corresponding ReachCode. The channel geometry can also be displayed in 3-D perspective using 3-D cross-section lines.

In Arc Hydro, cross-section points of a CrossSection are defined by x-, y-, z-, m-coordinates, where x and y locate a cross-section point on the ground, z represents the elevation of the point with respect to a datum, and m represents the linear measure of the point along the cross section. Using the m- and z-coordinates from each point, a transverse profile or geometry of a cross section can be drawn as a 2-D distance-elevation graph.

In the case of a traditionally surveyed cross section, where the cross-section geometry is stored as CrossSectionPoints, drawing a transverse profile requires three steps. First, the points are selected for a given CSCode. These points are then sorted on CrossM in ascending order. Finally, the CrossM and Elevation values are used to draw a 2-D distance-elevation graph. Whether or not the points need to be sorted depends on the type of graph used to plot the geometry.

If elevation values are not attached to the vertices of a profileline, drawing the longitudinal profile of a channel requires more work than a simple transverse profile. To draw a longitudinal profile of a thalweg, we first use ReachCode to select all the CrossSections that provide elevations. The lowest point of a cross section is assumed as the thalweg. The elevation of the thalweg position is read from the elevation (z-coordinate) field of the CrossSection 3-D line. Finally, the elevations are brought back to the profileline and plotted with the stations of cross sections. Longitudinal profiles for banks or other features can be drawn similarly. Bank and thalweg profiles

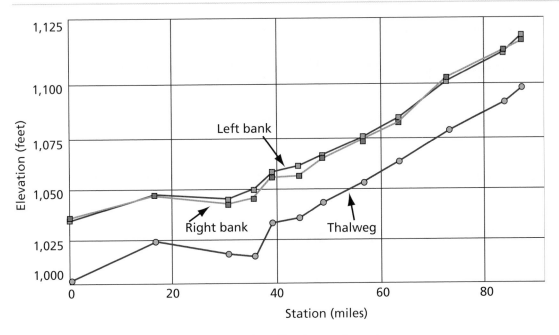

Longitudinal profiles along a channel

are very important for determining the flow pattern and reaches where river water spills into floodplains.

Every CrossSection 3-D line has a unique CSCode, and its properties in the event table are linked with it through this CSCode. The linear measure (m-coordinate) of the cross-section points is used to determine a position based on the measure value stored in the event table. By treating the CrossSection as a "route" and "event table," the property locations can be displayed as events of a cross section using linear referencing.

Creating a cross-section geodatabase

The example on the following page illustrates how all of the important channel information is integrated in a geodatabase using Arc Hydro.

Assume that a water resources consulting company is anticipating a project and decides to create a cross-section database using Arc Hydro. The company wants to concentrate on a portion of the river network system in the Waller Creek watershed in Austin, Texas. The river system includes upstream and downstream reaches of the main channel and a single reach representing a channel branch or tributary. The company has also collected a digital terrain model and a digital image of the watershed for this purpose.

As the GIS specialist of that company, you could follow these steps in order to complete your project. This example assumes that you have already created a geodatabase based on the Arc Hydro data model and are trying to put channel information into it. It also assumes that you have built-in tools at your disposal to perform each task.

— Upstream
— Downstream
— Branch

Waller Creek watershed river system

Exploring input data

A digital image of the Waller Creek watershed shows the alignment of the river network including banks, buildings, and other spatial features. This can be used as a background for digitizing profilelines and cross-section cutlines.

Cross-section geodatabase for Waller Creek watershed

The digital terrain model is a TIN that has been created from the interpretation of aerial photogrammetry using mass points and breaklines. The TIN represents the topography of the watershed and is the source of ground elevations for cross sections and profilelines. In this example, elevations for cross sections are extracted from the TIN.

Digital image of Waller Creek watershed

Digital terrain model (DTM) of Waller Creek watershed

Defining profilelines

The required profilelines that must be digitized are flowlines or thalwegs, banklines, and floodlines. To do this, you display the image in the background and digitize lines following the center of Main and Branch rivers. This task is best done by careful examination of a digital orthophoto of the channel. For simplicity, let us assume that the line you have digitized is the thalweg. Assign a unique ReachCode for each branch and its associated banks and floodlines.

The floodlines are intended to show where the water will flow when the channel overflows its banks. A simple way to locate them is to buffer the thalweg line by an arbitrary distance. However, in a river whose main channel has wide meanders through a broad river valley, the floodlines should follow the direction of the river valley and not the main channel.

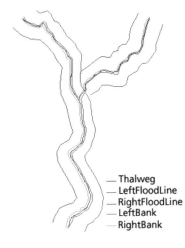

_Thalweg
_LeftFloodLine
_RightFloodLine
_LeftBank
_RightBank

Digitized ProfileLines

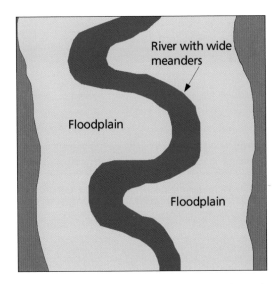

Floodlines follow the direction of the river valley in the case of a wide meandering river.

Defining cross-section cutlines

The purpose of the cross sections is to define channel geometry. You can digitize as many cross sections as needed. As a guideline, you can have widely spaced cross sections in a straight reach but may prefer to add more around bends to capture the changes in geometry. The flood-lines help to decide the extent of cross sections over floodplains. Cross sections should have a unique CSCode and the ReachCode from the branch they are representing. Once you have finished digitizing cross sections, the features in the database should look similar to the following cross section.

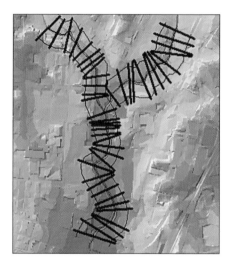

Cross-section cutlines and profilelines

Extracting elevations from the TIN

The ArcGIS 3D Analyst™ extension provides utilities to extract elevations from a surface using features. Specify the TIN as a surface and use cross-section cutlines and profilelines to extract ground elevations from the surface. The 3D Analyst tool assigns z-coordinate values to each vertex and thus converts two-dimensional lines into three-dimensional lines.

Assigning measures to linear features

In this step you add the linear measure value (m) at each vertex using linear referencing. You can either compute m-values from x,y coordinates or you can assign those values interactively. In this case of a cross section, the m-coordinates are consistent with its length and, therefore, it is easier to let the system compute m-values for you. However, if the actual length or stationing for rivers and embankments varies from the digitized length of the feature, assign m-coordinate values interactively at critical points.

Creating cross-section events

Once you have assigned measure values to cross sections, you can determine the position of any property point along the cross section. Or if you know the measure value of a property point, you can locate that position on the cross section using linear referencing. Now it is time to create cross-section events, which are similar to hydro events. You can create a point events table to store the location of the thalweg, banklines, and floodlines and a line event table to store properties like channel roughness. The CSCodes of the events should match the CSCodes of the cross sections they are on. The event tables may look similar to the following examples:

OID*	CSCode	MEASURE	TYPE
1	1	54.206	2
2	1	385.423	1
3	1	583.457	3
4	4	44.4	2
5	4	356.912	1
6	4	598.1	3
7	7	211.681	2
8	7	333.414	1

Example of a point event table

OID*	CSCode	FROMMEASUR	TOMEASURE	TYPE	CSPVALUE
1	1	0	116.05	0	0.01
2	1	116.05	583.46	0	0.002
3	1	583.46	781.5	0	0.01
4	4	0	130.78	0	0.01
5	4	130.78	533.02	0	0.002
6	4	533.02	720.23	0	0.01
7	7	0	211.68	0	0.01

Record: 5 ▶ ▶I Show: All Selected Records (0 out of 48 Selected.)

Example of a line event table

Retrieving data from the database

Finally, it is time to retrieve the channel information from the geodatabase. Using all these Cross-Sections, you can display the channel morphology by creating a perspective view in 3-D.

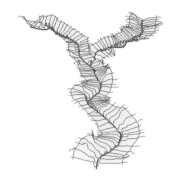

Three-dimensional view of Waller Creek river system

You can display the shape of a cross section or plot any profile along a profileline (thalwegs, banks, and floodlines) using measure and elevation values.

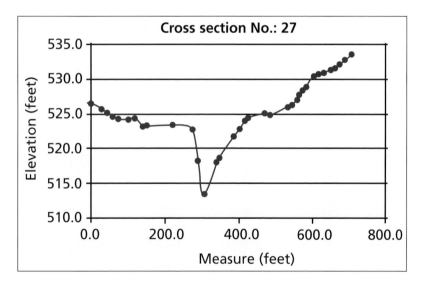

Geometry of a cross section

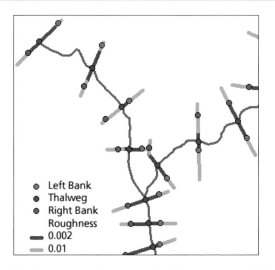

Displaying cross-section properties from an event table

By taking advantage of linear referencing and event tables, you can also display point and linear events along cross sections.

Other sources of cross-section data

So far we have discussed extracting cross sections from digital terrain models (DTMs) in a GIS or by surveying cross sections as a series of x,y,z coordinates using traditional methods or advanced surveying tools like GPS. But cross sections have been surveyed and used in water resources analysis for decades prior to the advent of computer-based GIS or even river hydraulic models. These cross sections are extremely valuable for the study of river morphology or for simulating historical events. This section briefly introduces some of these data sources and discusses how they can help to improve the channel shape representation extracted from DTMs.

Traditionally, cross sections were presented and stored in the form of drawings prepared by draftpersons based on survey information. Many water resources–related organizations have archived much historical information in this form. To bring the data into a model, the drawings must be transferred to distance-elevation formats. Transferring such data is time consuming and may not be cost effective for many projects if the corresponding stationing of the cross section along the river is only known approximately.

Another important source of cross-section data is the hydraulic models themselves. The cross-section information is stored either as an integrated part of a model or as an input file to a model. However, in both cases, cross-section geometries are most likely stored in a distance-elevation format with reach name and station as identifiers.

Channel	Main
Reach	Downstream
Station	171.58

Distance	Elevation
0.000	517.952
49.883	514.760
79.807	512.846
107.209	511.158
138.552	510.127
173.063	508.565
197.205	508.492
212.254	507.937
～～～	～～～
～～～	～～～

Cross-section data file prepared for a hydraulic model

The main advantage of cross sections from hydraulic models is that this cross-section information is in digital format. Using linear referencing and other utilities available in GIS, these cross sections can be imported into Arc Hydro with a reasonable amount of effort. It may be difficult, however, to reconcile this type of cross-section information with independently developed terrain surfaces.

Currently, the digital terrain models (DTMs) or digital elevation models (DEMs) available from different organizations usually depict the general land-surface topography of a watershed. Because of lower resolutions, these DTMs fail to provide detailed channel information, which is acceptable and probably desired for most river channel applications. However, a DTM can be

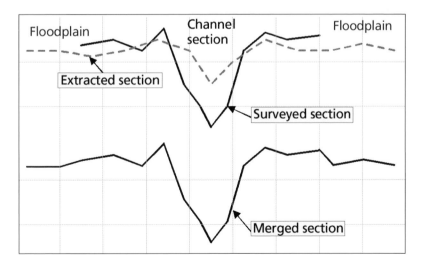

The combination of surveyed and extracted cross sections provides a better channel representation.

used for cross-section extraction because it provides sufficient information about the floodplain. A surveyed cross section represents the main channel shape better, but it typically does not cover a considerable length over the floodplain. These two data sources, if combined appropriately, can improve channel shape representation significantly.

What's next?

Streams and rivers are among the most fascinating and complex ecosystems on our planet. Their roles in providing natural resources, such as fish and clean water, are well known, as are their roles in providing transportation, energy, diffusion of wastes, and recreation. What is not as well known is how they serve as integrators of broader environmental conditions reflecting the surrounding landscape. Today, with ever-increasing demands being made on streams and rivers, the need to understand streams as ecological systems and to manage them effectively has become increasingly urgent.

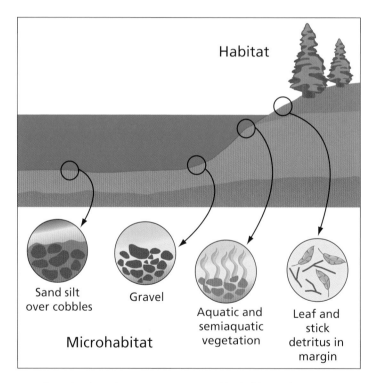

Organization of a stream system and its habitat subsystems
Source: Naiman, Robert J., and Robert E. Bilby (eds.) (1998), *River Ecology and Management,* Springer-Verlag, New York.

Human activities alter the physical habitats in rivers of all sizes. Extensive channelization and diking of large river systems for flood control and transportation are the primary causes for losing many secondary channels, backwaters, and oxbows, which are important habitats for many juvenile fish. Activities within a watershed, whether natural or human-induced, influence the most basic aspects of the hydrologic cycle, which in turn directly affect habitat distribution, trophic structure, physical and biological processes (such as sediment transport, nitrogen cycling, and primary production), and demography of the diverse biological communities. At a local habitat scale, substratum and current velocity are probably the most important factors determining the type of macroinvertebrate taxa present. The stream substratum has obvious importance because the vast majority of stream microinvertebrates spend most of their lives attached to substrata. The particle size of inorganic matter has a large influence on microinvertebrate community structure. For example, coarser bed materials (e.g., gravel, cobbles, and boulders) generally provide a more interstitial habitat for macroinvertebrates than fine sediments (e.g., sand, and silt).

Channel transects, depicting aquatic and semiaquatic vegetation conditions and particle-size distributions, along with river-flow and water-quality information, provide valuable insight about stream habitats and overall ecological conditions. In the field of ecology, rivers and streams serve as a circulatory water-flow system, and the study of those rivers, like the study of blood, can diagnose the health not only of the rivers themselves but also of their surrounding environments.

Arc Hydro provides an excellent means of capturing the three-dimensional nature of a channel. The data model can be extended to incorporate physical properties, such as the roughness of the channel bed, the nature of the vegetation and habitat conditions, and other important information essential for studying the complex nature of a river system.

Data dictionary

The diagram on the following page summarizes the object and feature classes in the Channel component of Arc Hydro, and their interrelationships. All the classes shown are available for loading data because they have inherited all the attributes from classes located above them in the UML hierarchy. The attributes shaded in blue are ESRI standard attributes, while those shaded in white are Arc Hydro attributes. The terms used in this diagram are defined in the glossary at the back of this book.

Simple feature class
CrossSection

Geometry *Polyline*
Contains M values *Yes*
Contains Z values *Yes*

The cross section of a channel, normally drawn transverse to the flow

Field name	Data type	Allow nulls	Default value	Domain	Precision	Scale	Length
OBJECTID	OID						
Shape	Geometry	Yes					
HydroID	Integer	Yes			0		
HydroCode	String	Yes					30
ReachCode	String	Yes					30
RiverCode	String	Yes					30
CSCode	String	Yes					30
JunctionID	Integer	Yes			0		
CSOrigin	String	Yes					30
ProfileM	Double	Yes			0	0	
Shape_Length	Double	Yes			0	0	

Unique identifier in the geodatabase
Permanent public identifier of the feature
An identifier for a river or stream segment
An identifier for a river
An identifier for a cross section
HydroID of the related HydroJunction
A classifier for the method by which the cross section was defined
The measure location of the cross section along the stream profile

Relationship class
CrossSectionHasPoint

Type *Simple*
Cardinality *One To Many*
Notification *None*

Forward label *CrossSectionPoint*
Backward label *CrossSection*

Origin feature class	Destination table
Name *CrossSection*	Name *CrossSectionPoint*
Primary key *CSCode*	
Foreign key *CSCode*	

No relationship rules defined.

Table
CrossSectionPoint

A point on the cross section

Field name	Data type	Allow nulls	Default value	Domain	Precision	Scale	Length
OBJECTID	OID						
CSCode	String	Yes					30
CrossM	Double	Yes			0	0	
Elevation	Double	Yes			0	0	

An identifier for a cross section
The measure location of the point along the cross section
Elevation of the point above mean sea level

Simple feature class
ProfileLine

Geometry *Polyline*
Contains M values *Yes*
Contains Z values *Yes*

Longitudinal profile of a stream or river channel

Field name	Data type	Allow nulls	Default value	Domain	Precision	Scale	Length
OBJECTID	OID						
Shape	Geometry	Yes					
HydroID	Integer	Yes			0		
HydroCode	String	Yes					30
ReachCode	String	Yes					30
RiverCode	String	Yes					30
FType	String	Yes					30
ProfOrigin	String	Yes					30
Shape_Length	Double	Yes			0	0	

Unique identifier in the geodatabase
Permanent public identifier of the feature
An identifier for a river or stream segment
An identifier for a river
A descriptor of feature type
A classifier for the method by which the profileline was defined

116

Hydrography

Kim Davis, University of Texas at Austin
Jordan Furnans, University of Texas at Austin
David Maidment, University of Texas at Austin
Victoria Samuels, University of Texas at Austin
Kristina Schneider, University of Texas at Austin

Hydrography is the map representation of surface-water features in the land-scape. Hydrography data sources include point, line, and area data layers from map hydrography, point features derived from tabular data inventories, and hydro response units that account for vertical exchanges of water in the hydro-logic cycle of a land surface. Arc Hydro was designed so that its features and attributes correspond with national and regional hydrography sets, including the National Hydrography Dataset (NHD).

Representing hydrographic features

Topographic maps contain various layers of information: political boundaries, cities, transportation routes, land-cover type (such as green for forest), hypsography (land-surface contours), and hydrography. The hydrography layers, usually colored blue, depict water features of the landscape, including streams, rivers, lakes, coastlines; and associated features such as swamps and wetlands; and water structures, such as bridges or locks on rivers.

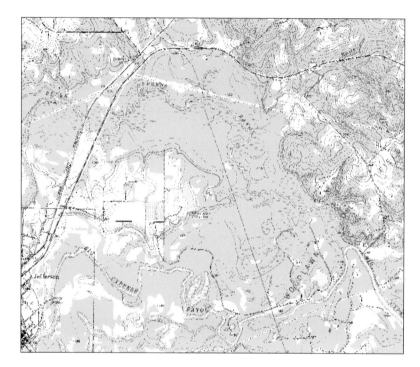

Hydrography on a topographic map

Map hydrography features can be classified in various ways, according to:
- dimensionality of the feature (point, line, or area)
- function of the feature (dam, stream, or pond)
- cartography of the feature (ephemeral stream, perennial stream, or double-line stream)

Different organizations classify features on a map according to one of these approaches. The Arc Hydro data model can accommodate all of these approaches by allowing features to have more than one classification. For example, features classified as points, lines, or areas may be further classified according to function and cartographic symbology. Arc Hydro uses the attribute FType as a generic text description of the feature type, and hydrography features are also classified into HydroPoint, HydroLine, and HydroArea classes. A given feature type, such as a dam, could be a point, a line, or an area, depending on the scale of map information used to represent it.

In addition to map hydrography, water feature information is also obtained from many tabular data sources. For example, in the United States the U.S. Army Corps of Engineers maintains a National Inventory of Dams that lists thousands of dams in the country. For each dam, an extensive list of attributes is maintained, including the latitude and longitude of the dam location. The U.S. Geological Survey publishes the latitude and longitude of its monitoring stations and the U.S. Environmental Protection Agency maintains a Permit Compliance System for water supply and wastewater discharge facilities. These information sources are important whenever hydrologists build a comprehensive water resources database for a region. In Arc Hydro the HydroPoint feature class has a series of subclasses intended to capture these sources of point information, including: Dam, Bridge, Structure, MonitoringPoint, WaterWithdrawal, WaterDischarge, and UserPoint.

One of the first things immediately apparent on a topographic map is a simple definition of the character of the land surface. Forested areas are shown as green, water areas as blue, cities as black, and so on. The way the landscape responds to rainfall depends a great deal on land cover, and to some degree on other factors, such as soil type. These factors may collectively be represented on a GIS map as hydro response units, a classification of the land surface into a set of polygons. Each hydro response unit is considered homogeneous in its vertical water-balance properties (i.e., the properties that define how precipitation will be transformed into runoff, evaporation, and groundwater recharge at a point on the land surface).

National and regional data sets

Water feature descriptions developed independently by various organizations have characteristics in common, such as connectivity of the flow network using centerlines through water bodies. Hydrography data models, too, have a great deal in common. However, they also vary, both in their approaches to feature descriptions and linear referencing, and in how the features are coded. Arc Hydro has been designed so that national and regional data sets can be imported into its data structure. Further, the hydro network and hydrography classes provide a generic set of classes upon which comparable hydrographic data layers can be constructed for regions that do not have them by customizing and extending the core Arc Hydro data model.

National Hydrography Dataset

The National Hydrography Dataset (NHD), developed by the USGS and EPA, is a data model for representation of map hydrography. It delivers a comprehensive hydrographic network data set of the United States at 1:100,000 scale, and is also used to create NHD data at larger scales, such as 1:24,000-scale data in some states. Stream hydrography in the NHD is described by 52 feature types that can be grouped into three main categories: drainage network features, water body features, and landmark features. Drainage network features describe single blue-line flow paths on the map (stream/river, canal/ditch, pipeline, and connector), as well as artificial paths through the center of water bodies and along coastlines. Water body features describe items such as lake/pond, sea/ocean, or 2-D stream/river, where the feature on the map is a blue-shaded area. Landmark features describe entities associated with the water system such as dam/weir, lock chambers, rapids, bridges, and gaging stations. The NHD feature representation provides a faithful reproduction in digital form of all the hydrographic features on USGS topographic maps.

To provide a clear description of the many hydrographic features in the NHD, two attributes are given to the point, line, and area features. The first feature-type descriptor, FType, names and describes the feature, such as stream/river or sea/ocean. Many feature types also have additional characteristics or properties of the features. The attribute FCode gives a numerical description of the feature type and the values of the characteristics. Using the FType and FCode attributes, the NHD gives the user both a general description of the feature type and a detailed numeric representation and corresponding characteristics, such as the FType = "Stream/River," and the FCode = "ephemeral or perennial."

Some NHD features have more than one possible spatial representation. For instance, dams can be areas or lines depending on map scale. This means dams can be represented in more than one Hydrography class and a relationship can be created between them if necessary.

Before the advent of the National Hydrography Dataset, several earlier versions of the United States river network were developed as the EPA River Reach Files. These files were created to provide river addressing systems, permit flow tracing through a river network, and attach flow attributes to the reaches, such as representative discharge and flow velocity. The NHD is cataloged and distributed using USGS eight-digit hydrologic cataloging units to define the spatial boundaries of each individual data set.

A typical river reach is a stretch of water between two confluences on a river. River reaches are defined in the NHD as sets of one or more drainage network features. Reaches are identified by a Reach Code, a 14-digit number comprised of the eight-digit HUC unit containing the reach and a six-digit segment number sequentially assigned within that HUC unit. Water body reaches are similarly defined and numbered based on lake/pond features. Coastline reaches are built on the artificial paths along coastlines, using water to the right to define the direction of tracing from one reach to the next. A Shoreline reach type has been defined in the NHD but not yet implemented.

The Reach Code is an identifier that tags each water feature uniquely within the drainage systems of the United States. The current publicly available version of the NHD, called NHDinArc, uses routes and regions in ARC/INFO® 7 to identify the line segments comprising a reach. The set of line segments making up a given NHD reach may be branched, such as the set of artificial paths forming the centerlines of a water body. Since network features in ArcInfo 8 can

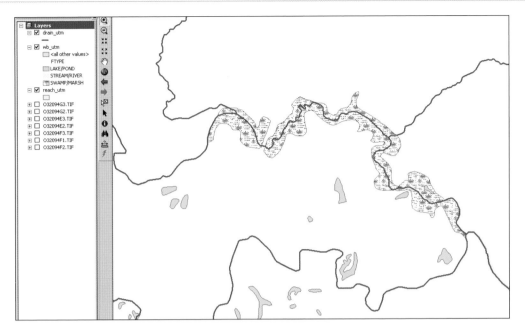

Hydrography from the National Hydrography Dataset (NHD)

only be linear, not branched, it is not possible to represent branched NHD reaches with a single network edge. For this reason NHD reaches are defined in Arc Hydro using linear referencing on sets of edges in the hydro network, rather than by associating each NHD reach with a single network edge.

The availability of the National Hydrography Dataset has been important for Arc Hydro for several reasons. First, it provides hydrographic data with which the Arc Hydro model can be applied at any location in the United States. Second, as a national hydrography data model design, it served as an important guide for the appropriate design of Arc Hydro. Third, its Reach Code description system provides the default system for water addressing in the United States, which will be used by the EPA for such tasks as filing permit requests for wastewater discharges. In fact, there are correspondences between the data structures.

Hydrography features in Arc Hydro
As shown in the UML diagram for the Hydrography feature data set, Feature is the parent class for HydroFeature, which is the parent class for Hydrography. A feature is any row in a table that has a unique identifier and a shape. Shapes can be 0-dimensional (points), 1-dimensional (lines), or 2-dimensional (areas). Arc Hydro provides a feature class for each dimension, HydroPoints, HydroLines, and HydroAreas. The HydroPoint class has seven child classes: MonitoringPoint, Dam, Bridge, Structure, WaterWithdrawal, WaterDischarge, and UserPoint. The UserPoint class is meant to hold point features that do not fit into the other classes. HydroArea contains one child class, Waterbody. The feature class HydroResponseUnit is included in Hydrography to

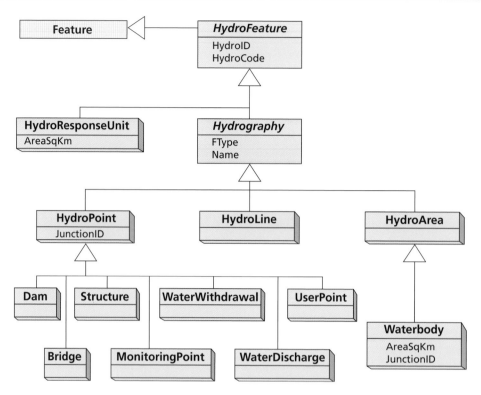

Hydrography representation in Arc Hydro

accommodate any type of response unit needed by a user. Each of these classes is discussed in greater detail in this chapter.

The following table shows the areas of correspondence between the data structures of the NHD and Arc Hydro:

National Hydrography Dataset	Arc Hydro Data Model
Drainage Network Features	Flowline HydroEdges, HydroJunctions
Artificial Paths for Coastlines	Shoreline HydroEdges
Water Bodies	Waterbody
Landmark Features	Hydro Points, Hydro Lines, Hydro Areas
Reach Code	ReachCode as reach identifier for linear referencing

Correspondence between data structures of the NHD and Arc Hydro

The authors of this book have considerable experience in building river basin networks, both before and after the release of the NHD, and the reduction in effort required to build a basin network using NHD compared to using the predecessor, River Reach File 3, is huge. There is no doubt that the advent of the NHD is a significant event in the evolution of GIS in water resources in the United States.

Since the NHD contains 52 different feature types with various characteristics, the question arises: how should these attributes be represented in Arc Hydro format? There are several levels of feature typing. The first level is just a simple string descriptor such as FType, the default already defined in Arc Hydro. The next stage of feature type description is to use a Coded Value Domain on FType. Subtyping a class of HydroPoint, HydroLine, or HydroArea is the third level of feature typing. The final and most descriptive phase of feature typing is subclassing a feature class. To convert the NHD to Arc Hydro, the Arc Hydro development team explored the feature typing possibilities available with subtyping and subclassing. We found that these methods created many unnecessary classes, in fact, more than in all the rest of Arc Hydro combined. The NHD development team at the U.S. Geological Survey and the EPA were active partners in the design of Arc Hydro, and they are developing a customized version of the Arc Hydro data model in which to publish the NHD.

Regional hydrography data sets

In addition to the NHD, hydrography data models have also been defined by several states and regions within the United States, and by organizations in other countries.

Questions concerning stream habitat for wildlife, access to spawning areas for fish, and the impact of forest logging make the Pacific Northwest a region where stream and river ecosystem management is a sensitive and relevant issue. The Washington Hydrography Framework is an effort in the state of Washington to develop and maintain a high-resolution hydrography data set. Coordinated with similar efforts in Oregon and northern California, a regional hydrography framework for the Pacific Northwest was created.

Hydrographic features in the Washington Hydrography Framework are modeled in four layers: Water Point, Watercourse, Water Body, and Shoreline. Water Points represent springs, seeps, and other point features. Watercourses represent streams, canals, flumes, pipelines, and centerlines through areal water features and double-line streams. Water Bodies include sounds, bays, lakes, ponds, wetlands, reservoirs, and inundation areas. Water Body Shorelines consist of one or more representations of a water body shoreline (e.g., mean high water, mean low water), one of which is chosen as the default shoreline and made coincident with the water body perimeter. Linear events are defined on the shoreline with the measure direction set, so that water is always on the right side of the line.

The Washington Hydrography Framework assigns to watercourses an LLID or Longitude/ Latitude Identification Number. The LLID is a 13-character concatenation of the longitude and latitude in decimal degrees of the location of the downstream mouth of the watercourse. Linear referencing may be measured by flow distance in kilometers upstream from this location.

The Wisconsin Department of Natural Resources has developed detailed data capture and feature coding decision rules for its 1:24,000 hydrography data layer. Two character feature codes are used to label various line, polygon, and associated features, such as secondary flows and bank features. During the last great glaciation of the earth, Wisconsin was covered by a

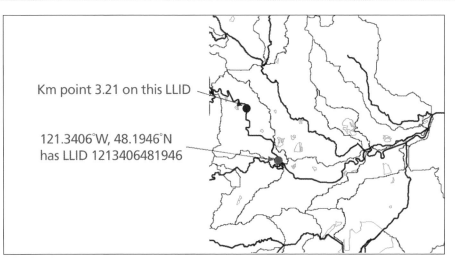

Km point 3.21 on this LLID

121.3406°W, 48.1946°N
has LLID 1213406481946

Linear referencing on watercourses in the Washington Hydrography Framework

vast ice sheet. The retreating ice left lake regions with complex drainage structures that constitute an ultimate challenge for hydrographic description. Lakes contain islands that themselves contain lakes!

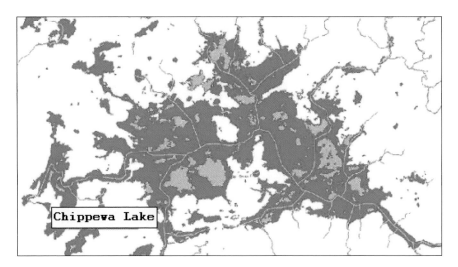

Chippewa Lake

Hydrography of Chippewa Lake, Wisconsin

The British Columbia Ministry of Fisheries has compiled a watershed atlas that inventories hydrographic features according to the watershed in which they fall. Watersheds are defined in a hierarchy such that watersheds of order "n" are contained within watersheds of order "n+1." This data model also uses closure lines to complete lake shorelines and to complete the coastline where double-line river mouths discharge to the ocean. Representation lines and connection lines are used to ensure connectivity in the hydraulic network. Feature codes are four-character integers and cover many complex hydrographic cases, similar to the capture rules of the Wisconsin Department of Natural Resources.

Hydrography derived from aerial photography

Many users do not need access to regional or national data sets, but need, instead, locally generated data. A typical example of this is a city using aerial photography to gather data, then digitizing its own stream network. In Arc Hydro the original line work from these types of data is entered into the Hydrography class HydroLine, then the centerlines of streams are converted to network types FlowLine and ShoreLine. The other digitized data, such as banklines or floodplain extent, would remain in the HydroLine class or be incorporated in the Arc Hydro channel model.

City of Austin aerial photography and digitized stream network

Relating hydrography to time series data

MonitoringPoint is the only Hydrography class that relates to TimeSeries objects in the standard Arc Hydro model, but other relationships can be added if necessary. TimeSeries objects are assembled as tables of data from gage readings and other observations made at a specific location. To relate the TimeSeries object and MonitoringPoint, the TimeSeries object contains the attribute FeatureID, which is the HydroID of the point feature. See chapter 7 for a more complete description of time series data.

Features that are represented by Dam and Structure points[1]

HydroPoints

The HydroPoint feature class has seven child classes: MonitoringPoint, Dam, Bridge, Structure, WaterWithdrawal, WaterDischarge, and UserPoint. These child classes can be relationally connected to junctions for use in the hydro network for analyses, or they can simply represent cartographic point locations. Points can represent locations on a river network that are vital when determining the actual flow of the network.

Dam, Bridge, and Structure points are intended to represent features, man-made or natural, that restrict or change the movement of water. The Structure class represents other hydraulic structures besides dams and bridges in the network that change the hydraulic properties of the flow through the network by their presence. Typical examples of such structures include detention ponds on small streams, levees designed to hold back floodwaters, and weirs. Structures can also be natural features like waterfalls if they have a significant effect on the hydraulic properties of the network. Typically, an irrigation turnout gate is not classified as a hydraulic structure, but as a WaterWithdrawal point. However, if the network of irrigation ditches is included in the network analysis, then the turnout structure would be a Structure point and would be modeled similarly to a valve on a pipe network. Structures may be either subtyped or subclassed, depending on the kind of data they store.

WaterWithdrawal and WaterDischarge represent points at which flow is removed or added to the stream network. Usually the corresponding flow data comes from permitting agencies that allow water rights and withdrawal permits or from environmental agencies that permit discharges into the network. These points are significant to network analyses that deal with the mass balance of water and pollutants in the network, and are used as flags in a network analysis.

MonitoringPoints store the locations of gages that measure water quantity or quality, including water-quality monitoring stations, stream-gage stations, rain-gage stations, and any other type of fixed-location data collection point. These points can be tied to the time series data collected at their locations, through the identifier field FeatureID. This allows the display of gage data in graphical format and comparison of gage readings at different locations. MonitoringPoints are well suited for subtyping, since most of them have similar attributes. Arc Hydro does not specify the attributes because of the wide variety of attribute data needed throughout the water resources community. For example, a stream-gaging station is described

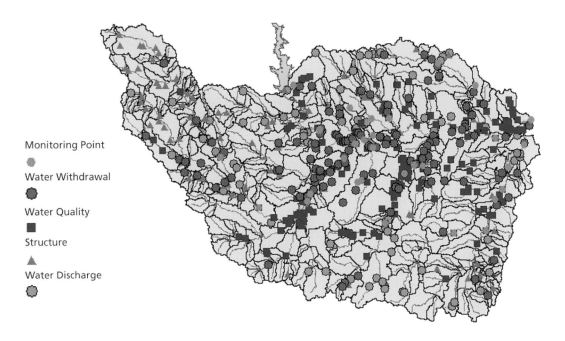

Monitoring Point

Water Withdrawal

Water Quality

Structure

Water Discharge

Point features and map hydrography

differently than a water-quality MonitoringPoint. More classes can be added to the data model to describe specific kinds of MonitoringPoints, if necessary.

UserPoints are intended to store point data that does not fit into the model elsewhere. These points might include locations where a river crosses an aquifer or a political or administrative boundary. UserPoints are a good place to load data that can be organized and exported to other classes after the Arc Hydro schema is applied.

Various water resource data inventories can be used with Arc Hydro, such as the National Inventory of Dams compiled by the U.S. Army Corp of Engineers and the Permit Compliance System (PCS) developed by the EPA to monitor the status of wastewater discharge permits. Many tabular data inventories have now been associated with geospatial data sets that represent their location, while others may contain latitude and longitude coordinate points that allow the data to be transformed to point features with attached attributes. The HydroPoint feature class in Arc Hydro is free of special attributes so that attributes of the tabular inventory from which the points are developed can be used.

HydroPoints store many specific types of points used in water resources analyses. However, they also can be used to store user points for purely cartographic purposes. These are points that serve to enrich the content of maps but are not used in analysis. Examples of such points are locations of isolated (off-channel) ponds, rock outcrops, small islands, and small springs.

Features that are represented by MonitoringPoints[2]

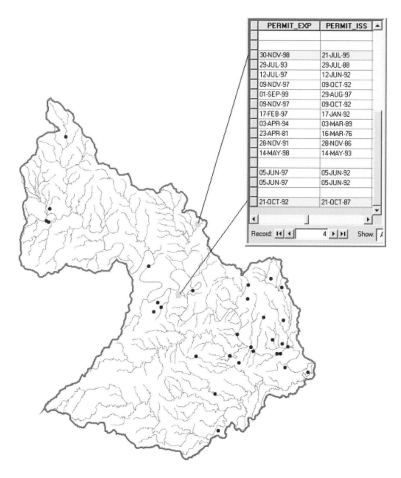

An example of data from the EPA Permit Compliance System inventory

HydroLines

HydroLines are designed to contain line features that are important for the cartographic representation of a water study area. Some examples of hydrographic lines are natural streams and rivers, man-made canals or ditches, pipelines that carry water underground, connectors that are used when the original data had some obstruction covering the hydrologic feature, and artificial paths that represent the centerlines of lakes and other water bodies. These features may also be represented in the hydro network.

Of course there are many more types of hydrographic lines participating in the network. Isolated ponds and lakes that are not part of the river network, shorelines, island boundaries, no-wake zones, swimming and recreation areas, roads, county and state boundary lines, jurisdictional boundaries for river authorities, and city limits are all marked off by lines that are important for cartography. HydroLines provide a spatial reference for viewers of the data and so are necessary in the model.

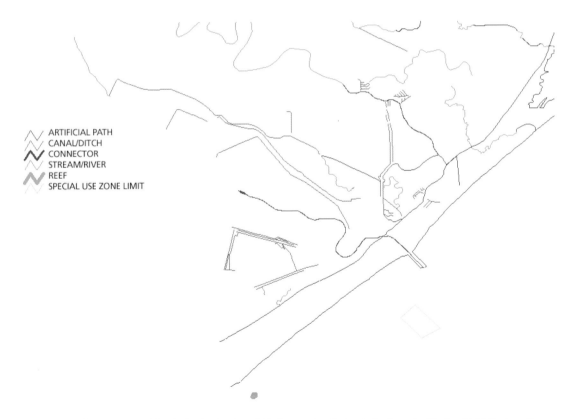

ARTIFICIAL PATH
CANAL/DITCH
CONNECTOR
STREAM/RIVER
REEF
SPECIAL USE ZONE LIMIT

HydroLines from a coastal basin in the National Hydrography Dataset displayed with the NHD Feature Type

HydroAreas

Ordinary landmark areas are stored as HydroAreas. Examples of these include no-wake zones within water bodies, extents of counties or other jurisdictional areas, and inundation areas.

Water bodies on a river channel and included in the network are located in the Waterbody child class of HydroArea. Land areas used in analysis, such as catchments and watersheds, land-use maps, and soil maps are stored in their respective classes as well, either Catchment, Watershed, or HydroResponseUnit. These classes are linked to the network through relationships with HydroJunctions so that the properties of these areas can be attached to the hydro network for analysis.

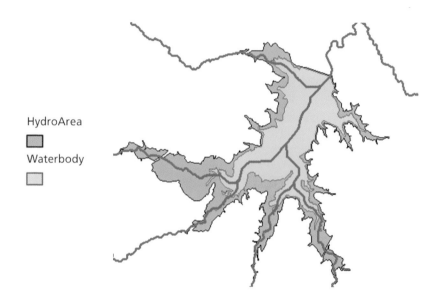

HydroArea is the area inundated when Benbrook Lake is at a high elevation.

HydroResponseUnits

A HydroResponseUnit is an area of the land surface that has homogeneous precipitation, land-surface characteristics, or both. HydroResponseUnits account for the vertical exchange of water through the hydrologic cycle and are used in hydrologic modeling simulations to accurately describe and predict how water will move through the environment.

Other HydroResponseUnits consider characteristics such as hydraulic conductivity, which is involved in the transfer of surface water to groundwater systems. Still other units are used in describing how precipitation is transformed to runoff on the land surface, and how this runoff migrates toward the ocean. Two general classes of HydroResponseUnits can be defined: units

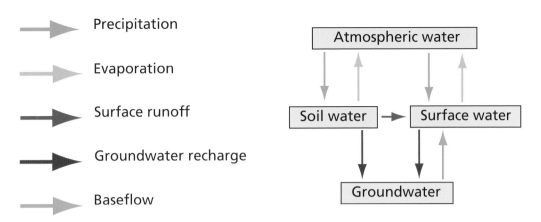

Precipitation

Evaporation

Surface runoff

Groundwater recharge

Baseflow

HydroResponseUnits allow for the vertical exchange of water in the hydrologic cycle

linking atmospheric and surface processes, and units linking surface processes and subsurface processes.

When defining hydro response units used to link atmospheric and surface processes that determine surface runoff from storm events, it is important to consider terrestrial characteristics such as land use and soil type. A standard runoff-prediction method, the Soil Conservation Service Curve Number method, considers both of these characteristics. In areas with high-porosity soils, infiltration is generally increased, which reduces the overall amount of surface runoff. Similarly, land use affects runoff by altering the permeability of the land surface. Urban areas with many buildings and much pavement have highly impervious surfaces and experience significant runoff. Agricultural areas, or areas of greater open space, however, tend to have less runoff for the same rainfall because the land surface is more receptive to infiltration.

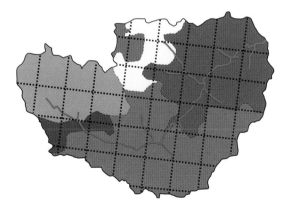

Hydro response unit cells (dashed boxes) overlain on drainage areas

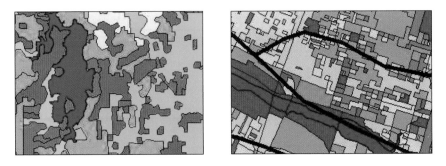

These maps show surface characteristics. The left map shows land classifications. The right map shows soil classifications from the Statsgo database for central Texas.[4]

The land-use and soils data for locations in Texas are shown in the maps above. In the map on the left, rangeland (brown) and forestland (green) are seen in proximity to a water body. This land-use coverage, obtainable through the U.S. EPA BASINS program[3], displays the non-uniformity in land use over relatively small areas. The soils data shown in the map on the right also displays great heterogeneity. The area shown is from an urbanized area in Austin, Texas, consisting of parkland (green) surrounding the river with the adjacent areas significantly covered with pavement. This data was obtained from the Statsgo database for central Texas.

Water is transferred between the atmosphere and the land surface through precipitation and evaporation. Each of these processes depends on the geographic location of the land area, and evaporation also depends on land use and season. Both processes are commonly recorded at specific points on the land surface, and methods have been determined to interpolate the respective values for the entire area based on only a few point measurements. One standard method is the use of Thiessen polygons. Using this method, the land surface is sectioned into polygons, and the data measured at one point is assumed to be representative of all the areas closer to that point than to any other point where measurements were made.

The land area is divided by intersecting the perpendicular bisectors of the lines connecting each of the measuring points. The outer boundary is often set at a certain distance from the points or at another known physical location. This process is can be done manually or with the

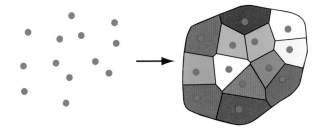

Thiessen polygon method for spatially distributing point measurements

Thiessen function in ArcGIS. An application using Thiessen polygons to determine catchment rainfall is presented in chapter 7.

Another common method of representing precipitation on a land surface is through Nexrad cells. These cells describe atmospheric moisture using remote-sensing technology. Data is available in grid format, as shown in the figure below. The Nexrad grid is intersected with a land-surface grid in order to determine the quantity of precipitation that has fallen onto a land area. Ingesting Nexrad time series data into Arc Hydro is shown in chapter 7.

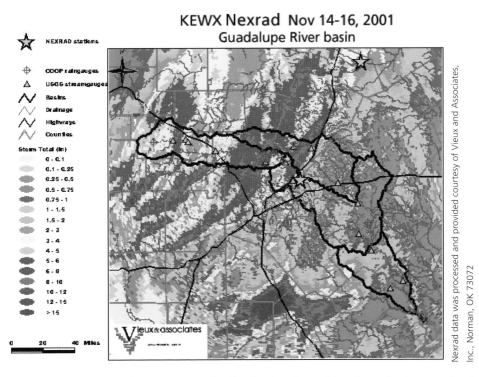

Nexrad data is available as a grid that is intersected with a land-surface grid to determine how much rain has fallen on the Guadalupe River basin.

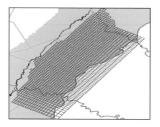

Groundwater modeling cells overlain by rivers and terrain boundaries

The second general type of hydro response unit, involved in subsurface-to-surface water exchanges is often represented by computational grids. These grids store numerous values for use in groundwater flow models, such as hydraulic conductivity, porosity, depth, and initial saturation. These properties are rarely homogeneous and isotropic over large subsurface areas. A popular groundwater flow model, MODFLOW, allows for the use of multilayered, irregularly spaced grids in its operation. Overlain on this grid are the geographical features of the study site, namely the rivers and property boundaries (green).

A particular challenge in linking surface and groundwater models is connecting the linear and areal features in surface water systems, defined by lines and polygons with the rectangular cell representation normally used for groundwater systems. This is an ongoing area of research.

Data dictionary

The diagrams on the following pages summarize the object and feature classes in the Hydrography component of Arc Hydro, and their interrelationships. All the classes shown are available for loading data because they have inherited all the attributes from classes located above them in the UML hierarchy. The attributes shaded in blue are ESRI standard attributes, while those shaded in white are Arc Hydro attributes. The terms used here are defined in the glossary at the back of this book.

Simple feature class
HydroPoint

Geometry *Point*
Contains M values *No*
Contains Z values *No*

A point feature on a map, such as a gage, well, or spring

Field name	Data type	Allow nulls	Default value	Domain	Prec- ision	Scale	Length	
OBJECTID	OID							
Shape	Geometry	Yes						
HydroID	Integer	Yes			0			Unique feature identifier in the geodatabase
HydroCode	String	Yes					30	Permanent public identifier of the feature
FType	String	Yes					30	Descriptor of feature type
Name	String	Yes					100	Geographic name
JunctionID	Integer	Yes			0			HydroID of the related HydroJunction

Simple feature class
Bridge

Geometry *Point*
Contains M values *No*
Contains Z values *No*

A structure where a road or railroad crosses a river or stream

Field name	Data type	Allow nulls	Default value	Domain	Prec- ision	Scale	Length	
OBJECTID	OID							
Shape	Geometry	Yes						
HydroID	Integer	Yes			0			Unique feature identifier in the geodatabase
HydroCode	String	Yes					30	Permanent public identifier of the feature
FType	String	Yes					30	Descriptor of feature type
Name	String	Yes					100	Geographic name
JunctionID	Integer	Yes			0			HydroID of the related HydroJunction

Simple feature class
Dam

Geometry *Point*
Contains M values *No*
Contains Z values *No*

An embankment or structure that ponds water to create a reservoir

Field name	Data type	Allow nulls	Default value	Domain	Prec- ision	Scale	Length	
OBJECTID	OID							
Shape	Geometry	Yes						
HydroID	Integer	Yes			0			Unique feature identifier in the geodatabase
HydroCode	String	Yes					30	Permanent public identifier of the feature
FType	String	Yes					30	Descriptor of feature type
Name	String	Yes					100	Geographic name
JunctionID	Integer	Yes			0			HydroID of the related HydroJunction

Simple feature class
Structure

Geometry *Point*
Contains M values *No*
Contains Z values *No*

A water structure not represented as a bridge or dam, such as a wier or a waterfall

Field name	Data type	Allow nulls	Default value	Domain	Prec- ision	Scale	Length	
OBJECTID	OID							
Shape	Geometry	Yes						
HydroID	Integer	Yes			0			Unique feature identifier in the geodatabase
HydroCode	String	Yes					30	Permanent public identifier of the feature
FType	String	Yes					30	Descriptor of feature type
Name	String	Yes					100	Geographic name
JunctionID	Integer	Yes			0			HydroID of the related HydroJunction

Simple feature class
UserPoint

Geometry *Point*
Contains M values *No*
Contains Z values *No*

A user-selected point location, such as where a river crosses an aquifer boundary

Field name	Data type	Allow nulls	Default value	Domain	Precision	Scale	Length	
OBJECTID	OID							
Shape	Geometry	Yes						
HydroID	Integer	Yes			0			Unique feature identifier in the geodatabase
HydroCode	String	Yes					30	Permanent public identifier of the feature
FType	String	Yes					30	Descriptor of feature type
Name	String	Yes					100	Geographic name
JunctionID	Integer	Yes			0			HydroID of the related HydroJunction

Simple feature class
WaterDischarge

Geometry *Point*
Contains M values *No*
Contains Z values *No*

A location where water is discharged to a river, stream, or water body

Field name	Data type	Allow nulls	Default value	Domain	Precision	Scale	Length	
OBJECTID	OID							
Shape	Geometry	Yes						
HydroID	Integer	Yes			0			Unique feature identifier in the geodatabase
HydroCode	String	Yes					30	Permanent public identifier of the feature
FType	String	Yes					30	Descriptor of feature type
Name	String	Yes					100	Geographic name
JunctionID	Integer	Yes			0			HydroID of the related HydroJunction

Simple feature class
WaterWithdrawal

Geometry *Point*
Contains M values *No*
Contains Z values *No*

A location where water is withdrawn from a river, stream or water body

Field name	Data type	Allow nulls	Default value	Domain	Precision	Scale	Length	
OBJECTID	OID							
Shape	Geometry	Yes						
HydroID	Integer	Yes			0			Unique feature identifier in the geodatabase
HydroCode	String	Yes					30	Permanent public identifier of the feature
FType	String	Yes					30	Descriptor of feature type
Name	String	Yes					100	Geographic name
JunctionID	Integer	Yes			0			HydroID of the related HydroJunction

Simple feature class
HydroResponseUnit

Geometry *Polygon*
Contains M values *No*
Contains Z values *No*

An area of the landscape possessing uniform properties for conversion of rainfall into runoff

Field name	Data type	Allow nulls	Default value	Domain	Precision	Scale	Length	
OBJECTID	OID							
Shape	Geometry	Yes						
HydroID	Integer	Yes			0			Unique feature identifier in the geodatabase
HydroCode	String	Yes					30	Permanent public identifier of the feature
AreaSqKm	Double	Yes			0	0		Area in square kilometers
Shape_Length	Double	Yes			0	0		
Shape_Area	Double	Yes			0	0		

See network features

Simple junction feature class
HydroJunction

Field name	Data type
OBJECTID	OID
Shape	Geometry
AncillaryRole	Small Integer
Enabled	Small Integer
HydroID	Integer
HydroCode	String
NextDownID	Integer
LengthDown	Double
DrainArea	Double
FType	String

Relationship class
HydroJunctionHasMonitoringPoint

Type *Simple*   Forward label *MonitoringPoint*
Cardinality *One To Many*   Backward label *HydroJunction*
Notification *None*

Origin feature class	Destination feature class
Name *HydroJunction*	Name *MonitoringPoint*
Primary key *HydroID*	
Foreign key *JunctionID*	

No relationship rules defined

Simple feature class
MonitoringPoint

Geometry *Point*
Contains M values *No*
Contains Z values *No*

Field name	Data type	Allow nulls	Default value	Domain	Prec-ision	Scale	Length
OBJECTID	OID						
Shape	Geometry	Yes					
HydroID	Integer	Yes			0		
HydroCode	String	Yes					30
FType	String	Yes					30
Name	String	Yes					100
JunctionID	Integer	Yes			0		

A location where water flow or properties are measured, such as a stream gage, rainfall gage or water quality monitoring site

Unique feature identifier in the geodatabase
Permanent public identifier of the feature
Descriptor of feature type
Geographic name
HydroID of the related HydroJunction

Relationship class
MonitoringPointHasTimeSeries

Type *Simple*   Forward label *TimeSeries*
Cardinality *One To Many*   Backward label *MonitoringPoint*
Notification *None*

Origin feature class	Destination feature class
Name *MonitoringPoint*	Name *TimeSeries*
Primary key *HydroID*	
Foreign key *FeatureID*	

No relationship rules defined

See time series objects

Table
TimeSeries

Field name	Data type
OBJECTID	OID
FeatureID	Integer
TSTypeID	Integer
TSDateTime	Date
TSValue	Double

137

Simple feature class
HydroLine

Geometry *Polyline*
Contains M values *No*
Contains Z values *No*

Field name	Data type	Allow nulls	Default value	Domain	Precision	Scale	Length
OBJECTID	OID						
Shape	Geometry	Yes					
HydroID	Integer	Yes			0		
HydroCode	String	Yes					30
FType	String	Yes					30
Name	String	Yes					100
Shape_Length	Double	Yes			0	0	

A line hydrography feature not represented by a HydroEdge, such as an administrative boundary

Unique feature identifier in the geodatabase
Permanent public identifier of the feature
Descriptor of feature type
Geographic name

Simple feature class
HydroArea

Geometry *Polygon*
Contains M values *No*
Contains Z values *No*

Field name	Data type	Allow nulls	Default value	Domain	Precision	Scale	Length
OBJECTID	OID						
Shape	Geometry	Yes					
HydroID	Integer	Yes			0		
HydroCode	String	Yes					30
FType	String	Yes					30
Name	String	Yes					100
Shape_Length	Double	Yes			0	0	
Shape_Area	Double	Yes			0	0	

An areal hydrography feature not represented by a Waterbody, such as a swamp or inundation area

Unique feature identifier in the geodatabase
Permanent public identifier of the feature
Descriptor of feature type
Geographic name

See network features

Simple junction feature class
HydroJunction

Field name	Data type
OBJECTID	OID
Shape	Geometry
AncillaryRole	Small Integer
Enabled	Small Integer
HydroID	Integer
HydroCode	String
NextDownID	Integer
LengthDown	Double
DrainArea	Double
FType	String

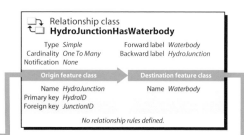

Relationship class
HydroJunctionHasWaterbody

Type *Simple*   Forward label *Waterbody*
Cardinality *One To Many*   Backward label *HydroJunction*
Notification *None*

Origin feature class	Destination feature class
Name *HydroJunction*	Name *Waterbody*
Primary key *HydroID*	
Foreign key *JunctionID*	

No relationship rules defined.

Simple feature class
Waterbody

Geometry *Polygon*
Contains M values *No*
Contains Z values *No*

Field name	Data type	Allow nulls	Default value	Domain	Precision	Scale	Length
OBJECTID	OID						
Shape	Geometry	Yes					
HydroID	Integer	Yes			0		
HydroCode	String	Yes					30
FType	String	Yes					30
Name	String	Yes					100
AreaSqKm	Double	Yes			0	0	
JunctionID	Integer	Yes			0		
Shape_Length	Double	Yes			0	0	
Shape_Area	Double	Yes			0	0	

Waterbodies are all the significant ponds, lakes, and bays in the water system.

Unique feature identifier in the geodatabase
Permanent public identifier of the feature
Descriptor of feature type
Geographic name
Area in square kilometers
HydroID of the HydroJunction at water body outlet

138

Data acknowledgments

[1] Dam and structure point pictures are from www.lcra.org/water/images/manflda.jpg and www.shogun.co.uk/watercpi/cpiphoto.htm

[2] Monitoring point pictures are from www.conantcustombrass.com/conant/vrg-2.html

[3] Data for the Land coverage HydroResponseUnit example is from www.epa.gov/OST/BASINS

[4] Data for Statsgo HydroResponseUnit example is from www.gis.uiuc.edu/nrcs/statsgo_inf.html

[5] Data for NEXRAD HydroResponseUnit example is from www4.ncdc.noaa.gov/cgi-win/wwcgi.dll? WWNEXRAD

Time Series

David Maidment, University of Texas at Austin
Venkatesh Merwade, University of Texas at Austin
Tim Whiteaker, University of Texas at Austin
Michael Blongewicz, DHI Water and Environment
David Arctur, ESRI

The flow and quality of water are defined by time series measurements taken at gages and sampling points. Some data is archived at regular intervals, such as daily mean discharge or daily precipitation, and other data is measured irregularly in time, such as water-quality samples taken a few times a year at a particular location on a stream. The synthesis of spatial and time series data is a particular challenge for Arc Hydro because GIS data models do not normally consider temporal information.

Time series data sources

Time series data archives such as the USGS National Water Information System or the EPA's Storet system for water-quality data have a complicated tabular structure. Arc Hydro does not try to duplicate this structure, but instead provides a repository for time series data derived from water measurements or simulation models. When daily streamflow data is downloaded from the USGS National Water Information System archives, the result typically appears in text form as a header sequence, followed by a sequence of four columns giving the location (station), time of measurement (datetime), type of data, and value, which is daily mean discharge of the river at this location.

```
#  U.S. Geological Survey
#  National Water Information System
#  Retrieved: 2002-05-09 19:44:37 EDT
#  This file contains published daily mean streamflow data.
#  This information includes the following fields:
#      agency_cd      Agency Code
#      site_no        USGS station number
#      dv_dt          date of daily mean streamflow
#      dv_va          daily mean streamflow value, in cubic-feet per-second
#      dv_cd          daily mean streamflow value qualification code
#
#  Sites in this file include:
#      USGS 08176500 Guadalupe Rv at Victoria, TX
#
agency_cd      site_no dv_dt dv_va dv_cd
5s 15s 10d      12n      3s
USGS   08176500         1998-10-01      869
USGS   08176500         1998-10-02      1130
USGS   08176500         1998-10-03      1050
USGS   08176500         1998-10-04      1030
USGS   08176500         1998-10-05      945
USGS   08176500         1998-10-06      1500      e
USGS   08176500         1998-10-07      3000      e
USGS   08176500         1998-10-08      6110
USGS   08176500         1998-10-09      6660
USGS   08176500         1998-10-10      3670
USGS   08176500         1998-10-11      1940
USGS   08176500         1998-10-12      1520
USGS   08176500         1998-10-13      1170
USGS   08176500         1998-10-14      1180
```

This text-format USGS time series data file contains records of daily mean river discharge on the Guadalupe River at Victoria, Texas.

The (time, value) pair is unique for each record, while the location and type information is the same for all records. This leads to the concept of a time series collection, which is a set of (time, value) pairs through a time period, the duration of the collection.

National Water Information System
The USGS National Water Information System (NWIS) provides online access to water resources data collected at USGS gaging stations in all 50 U.S. states, the District of Columbia, and Puerto Rico. NWIS data includes real-time and historical surface-water, groundwater, and water-quality data, including descriptive site information. The real-time data typically is recorded at 15- to 60-minute intervals, stored on-site, then transmitted to USGS offices every four hours via satellite, telephone, or radio.

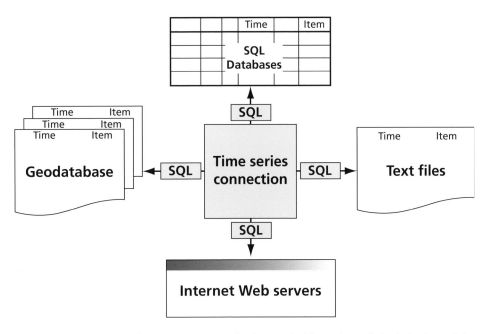

Data can be downloaded over the Internet; it can also be queried from dynamic hydrologic models running in parallel with the GIS, from other relational databases, from text files, or from files stored within the geodatabase itself.

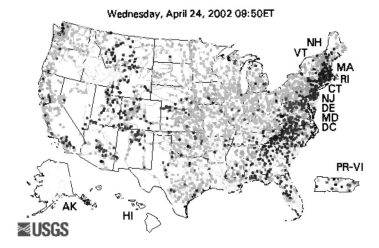

The USGS has approximately five thousand gaging stations across the United States reporting data in real time. Results are displayed as a continually updated color map. Green dots indicate stations where flow conditions are normal, blue dots show higher than normal flow conditions, and red dots show lower than normal flow conditions. The red dots on this map give evidence of drought conditions in the eastern states in the spring of 2002.

NWIS data can be viewed either as a graph or a table:

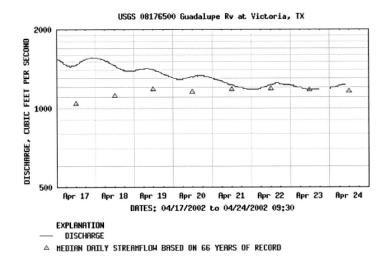

Graphical view of USGS real-time streamflow data

Since all NWIS data is available online, it is possible to extract it using a customized application that interacts with the NWIS Web site and the geodatabase. Temporal data can then be linked with the geospatial data in Arc Hydro, thereby creating a spatial and temporal framework that can be used for hydrologic modeling. For example, gaging stations are MonitoringPoints with time series for each point stored in the TimeSeries table. The time series data and the spatial features are related using the FeatureID of the TimeSeries table and the HydroID of the MonitoringPoint feature class.

Guadalupe basin with streamflow gaging station locations, stored as MonitoringPoints in Arc Hydro

The online NWIS data can be downloaded and stored in the TimeSeries table of Arc Hydro.

OBJECTID*	FeatureID*	TSTypeID*	TSDateTime	TSValue
485615	12000033	1	10/23/2000	541
485616	12000033	1	10/24/2000	486
485617	12000033	1	10/25/2000	682
485618	12000033	1	10/26/2000	923
485619	12000033	1	10/27/2000	723
485620	12000033	1	10/28/2000	892
485621	12000033	1	10/29/2000	705
485622	12000033	1	10/30/2000	595

Record: 1 — Show: All | Selected — Records (24046 out of 382417 Selected.)

An Arc Hydro TimeSeries table of daily mean streamflow records obtained for a gaging station at Victoria, Texas. A customized automated procedure to download NWIS data and store it in the Arc Hydro format is explained later in this chapter. The procedure was used to download the entire streamflow history of all 29 stations in the Guadalupe basin, over one thousand station-years of streamflow data. This generated a TimeSeries table with about 380,000 records, occupying about 20MB of disk space in a Microsoft Access database. A TimeSeries table stores a comprehensive inventory of time series information.

The two factors determining how time series data can be used are the length of the time series record and the time interval. Depending on requirements, hydrologists vary the frequency of recording. For example, more data is collected of the continuously changing hydrology of a catchment during a flood or during the rainy season. Hydrologists also need to know whether the data contains actual recorded values or interpolated values between two recordings, as this determines how other statistics are calculated. Hydrologic data is often incomplete because of recording devices that malfunction or because of manual error. Understanding all of these factors is important in understanding time series data and its use in hydrology.

Time series in hydrology and GIS

The intention of including time series data in Arc Hydro is not only to build a complete hydrologic data model for use within the ArcGIS environment, but also to create a database that is accessible to many water resources models that operate independently of the GIS. Hydrologic models have utilized time series data for decades. However, the problem of how to access the time series data stored in the GIS is still an issue for hydrologic models that work independently of GIS. This problem can be solved by integrating hydrologic models with the GIS database.

It is easier to conceive how a hydrologic model and the GIS could share a river network as a series of connected linear features, or how they might share catchment and watershed data as a series of adjacent polygons, than how the models could share time series data. To integrate GIS with hydrologic models, users need to know how the time series data is stored in GIS and also the format of the hydrologic/hydraulic model input files. It is also important to know how easy it is for a hydrologic model to read the time series data directly from the geodatabase. Successfully integrating GIS and hydrologic models depends on how the hydrologic models interact with the time series data stored in the GIS.

Other problems result from the fact that time series data is often voluminous; storing as well as accessing this data is potentially difficult. A single time series file typically holds more than 10 years of daily or hourly records made at irregular intervals. Historically, standard relational databases have been inefficient for retrieving this amount of data, as well as knowing the format of various types of data. Consequently, hydrologic models usually use their own formatted data files. This has led to the development of proprietary software using proprietary data files, which is contrary to the intention of Arc Hydro. Recently, however, large database vendors including Oracle and Sybase® have added time series capabilities to their applications.

One of the most typical functions of a time series database is to compile monthly and annual values from daily values. This seemingly simple task is complicated by the varying number of days in a month, the presence of leap years, and the existence of water years, which are specially defined 12-month periods that are more appropriate than calendar years for describing the annual cycle of water storage and depletion. For example, the U.S. Geological Survey stores its streamflow data in water years beginning in October and running through the following September, in order to better account for the cycle of snow accumulation and melting in mountainous regions. Since many hydrologic models use time steps of days or even hours, simulation of years of hydrologic conditions quickly creates voluminous data series. For many purposes, it is not necessary to carry out analyses with this degree of temporal analysis. Many hydrologic studies can be undertaken conveniently with mean annual or monthly data, or with data for particular extreme events, such as floods.

Time series in Arc Hydro

Time series data is captured and stored in a variety of formats. An Arc Hydro user may acquire, store, or deliver an entire hydrologic data set, including time series data files from monitoring and catchment stations.

All the examples presented in this chapter illustrate Arc Hydro time series as a large table of data containing all types of time series data for all features and for all times. However, to accommodate large data sets and many variables such as water-quality data, climatic data, and so on, the core model can be extended to have separate TimeSeries tables for different variables.

Variables, space, and time
Time series information can be depicted in 3-D space, where the three coordinate axes are space, indexed by location L; time, indexed by T; and the variable being measured, indexed by V. So the data value D can be written as $D(L,T,V)$ to symbolize its dependence on these coordinate axes.

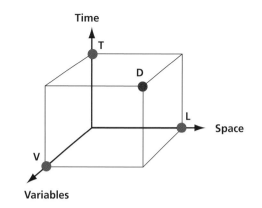

Data for a variable measured at a particular point in space and time

In a relational database, field names are used to represent variables, and those chosen for Arc Hydro are FeatureID to symbolize the spatial feature or location (L), TSDateTime to represent time (T), and TSTypeID to represent the variable measured (V). Thus, any measured value (TSValue, in this case) can be represented by a point in three-dimensional space, with its corresponding FeatureID, TSDateTime, and TSTypeID attributes.

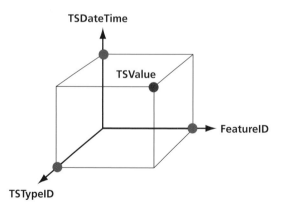

Three-dimensional structure of time series in Arc Hydro

This three-dimensional structure of the time series is simple and general. The structure is formed by a series of vertical and horizontal planes. Therefore, any TimeSeries table that has a single value for FeatureID represents a vertical plane perpendicular to the FeatureID axis, and it contains several values for TSTypeID and TSDateTime. A TimeSeries table that has a single value for TSTypeID represents another vertical plane that is perpendicular to the TSTypeID axis.

Similarly, a TimeSeries table that has a single value of TSDateTime represents a horizontal plane perpendicular to the TSDateTime axis. In this way different time series "views" for individual features and attributes can be created from the time series object class. These views are created by using a simple query definition in ArcGIS. When two vertical planes intersect, their line of intersection represents a time series view that corresponds to a single FeatureID and single TSTypeID for several TSDateTime values. The intersection of all three planes represents a single point that has only one value for FeatureID, TSTypeID, and TSDateTime, respectively.

Extracting time series views from the TimeSeries table

This section describes how different TimeSeries tables can be extracted from the time series data table. The data used here is daily streamflow and gage-height data for 18 gaging stations in the Guadalupe basin for the year 1999 only.

Suppose we want to select all the streamflow and gage-height data measured at the Guadalupe River gaging station at Victoria, which has a FeatureID of 12000033. This data view, represented by a vertical plane in the three-dimensional time series cube model, is indexed by the value of the FeatureID.

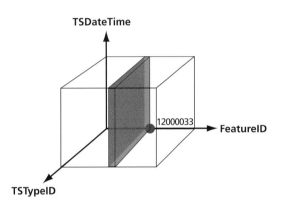

Extraction of time series for a given feature

So in order to extract time series for FeatureID 12000033, an attribute selection query can be performed in ArcMap as [FeatureID] = 12000033. The output table has only one FeatureID, and for this feature it contains time series data for two different types of TSTypeIDs, namely 1, streamflow and 2, gage height. Since the TimeSeries table contains daily data for 1999 only, 730 records are selected, 365 each for streamflow and gage-height, respectively.

OBJECTID	FEATUREID	TSTYPEID	TSDATETIME	TSVALUE
485400	12000033	1	12/26/1999	631
485401	12000033	1	12/27/1999	573
485402	12000033	1	12/28/1999	518
485403	12000033	1	12/29/1999	495
485404	12000033	1	12/30/1999	505
485405	12000033	1	12/31/1999	626
509559	12000033	2	1/1/1999	10.82
509560	12000033	2	1/2/1999	9.8
509561	12000033	2	1/3/1999	9.64
509562	12000033	2	1/4/1999	10.01
509563	12000033	2	1/5/1999	9.61

Record: 1 — Show: All Selected Records (730 out of 12716 Selected.)

Streamflow (TSTYpeID=1) and gage height (TSTypeID=2) data for the Guadalupe River at Victoria during 1999

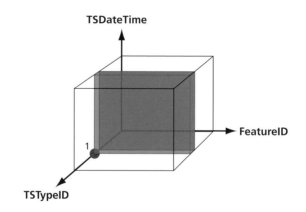

Extraction of time series for a given TSTypeID

Suppose instead we want to extract the streamflow discharge data for all gaging stations in the Guadalupe basin. This means selected data for a particular TSTypeID, stream flow in this case.

The attribute query on the TimeSeries table for this view is [TSTYPEID]=1. The result contains time series data for only one TSTypeID, 1, for all gaging stations for 1999.

OBJECTID	FEATUREID	TSTYPEID	TSDATETIME	TSVALUE
137958	12000001	1	4/19/1999	27
137959	12000001	1	4/20/1999	27
137960	12000001	1	4/21/1999	27
137961	12000001	1	4/22/1999	27
137962	12000001	1	4/23/1999	27
137963	12000001	1	4/24/1999	27
137964	12000001	1	4/25/1999	31
137965	12000001	1	4/26/1999	36
137966	12000001	1	4/27/1999	32
137967	12000001	1	4/28/1999	29
137968	12000001	1	4/29/1999	28

Record: 22 Show: All Selected Records (6358 out of 12716 Selected.)

Time series for a given TSTypeID and several features. Since the table contains data for only two TSTypeIDs, half of the data is selected as shown in this table.

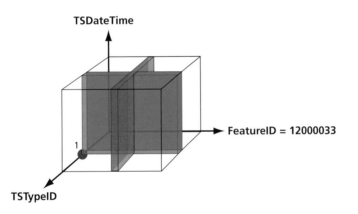

Extraction of time series for a single feature and a single TSTypeID

If we combine the two queries just executed, the result is a data view of a time series for one feature and one variable.

The blue plane represents the time series data for a given feature (Guadalupe River at Victoria) and the red plane represents the data for a particular TSTypeID (1 = streamflow discharge). The intersection between these two planes represents time series data that has a single value for FeatureID and a single TSTypeID. The attribute query for this table is [FEATUREID] = 12000033 AND [TSTYPEID] = 1 which results in 365 daily streamflow records being selected.

151

OBJECTID	FEATUREID	TSTYPEID	TSDATETIME	TSVALUE
485041	12000033	1	1/1/1999	2810
485042	12000033	1	1/2/1999	2660
485043	12000033	1	1/3/1999	3310
485044	12000033	1	1/4/1999	3260
485045	12000033	1	1/5/1999	2940
485046	12000033	1	1/6/1999	2640
485047	12000033	1	1/7/1999	2460
485048	12000033	1	1/8/1999	2370
485049	12000033	1	1/9/1999	2340
485050	12000033	1	1/10/1999	2330

Selected Attributes of TimeSeries

Record: 362   Show: All | Selected   Records   (365 out of 12716 Selected.)

Selected TimeSeries data table for a single feature and TSTypeID

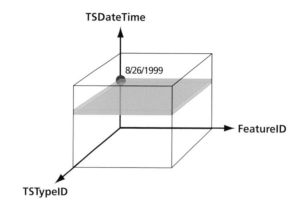

Extraction of time series data for a given TSDateTime

Finally, suppose we want to view all data recorded at all gaging stations at a point in time. For a given TSDateTime, a horizontal plane in the three-dimensional figure represents a data set at a particular point in time, which is like a normal feature attribute table in ArcGIS.

The query for this table is [TSDATETIME] = #8/26/1999#. It gives TSValues for all features and all TSTypeIDs at any given time, 8/26/1999 in this case.

OBJECTID	FEATUREID	TSTYPEID	TSDATETIME	TSVALUE
324537	12000017	1	8/26/1999	126
351074	12000019	1	8/26/1999	310
403594	12000022	1	8/26/1999	140
415132	12000024	1	8/26/1999	523
439236	12000026	1	8/26/1999	6.9
455948	12000030	1	8/26/1999	3.4
485278	12000033	1	8/26/1999	762
503825	12000035	1	8/26/1999	4.5
504128	12000001	2	8/26/1999	6.9
504425	12000002	2	8/26/1999	5.7

Record: 1   Show: All Selected   Records (36 out of 12716 Selected.)

Corresponding time series data for a given TSDateTime

Any hydrologic model or Microsoft Excel spreadsheet can directly interact with this database and extract the data, even without going through ArcGIS, because the TimeSeries table is a regular table in a relational database.

TimeSeries object class

The following UML diagram shows the core model for TimeSeries in Arc Hydro.

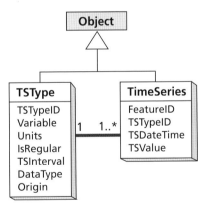

UML diagram for time series in Arc Hydro

The TimeSeries object class in the Arc Hydro data model has four attributes: FeatureID, TSTypeID, TSDateTime, and TSValue.

153

The FeatureID is an integer identifier, usually set equal to the HydroID of the feature described by the time series. For example if the HydroID of the MonitoringPoint feature is 12000033, then all the TimeSeries objects of this MonitoringPoint feature have a FeatureID of 12000033.

Each element in a TimeSeries class is classified as a distinct type of time series information, such as precipitation, streamflow, or evaporation. TSTypeID is an integer identifier that points to a specific record in the TSType table. This record contains descriptive information about the type of time series under which the associated TimeSeries object is classified.

Each element in the TimeSeries table has a distinct time stamp, or a labeled point in time. This time stamp is stored in the TSDateTime field in the TimeSeries table. This field is a standard Date field type in ArcGIS, which includes both date and time information. A TSDateTime value has the following format:

MM/DD/YYYY hh:mm:ss.sss
where:
MM is the month (01 to 12)
DD is the day (01 to 31)
YYYY is the four-digit year
hh is the hour (00 to 23)
mm is the minute (00 to 59)
ss.sss is the second with milliseconds (00.000 to 59.999)

Date fields in ArcGIS store date information down to the millisecond, although date values are displayed according to the precision of the value. For example, the TSDateTime value of 01/31/2001 12:00:00 A.M. would be displayed in ArcGIS as 01/31/2001. However, 01/31/2001 12:00:01 A.M. would be displayed as 01/31/2001 12:00:01 A.M. TSValue is the actual value recorded for this FeatureID, TSTypeID, and TSDateTime. Where time series data has been drawn from external databases, it is useful to also store the HydroCode of the geographic feature so that the time series data values can be traced to their point of origin if necessary.

TSType object class

The TSType object class in the Arc Hydro data model has seven attributes: TSTypeID, Variable, Units, IsRegular, TSInterval, DataType, and Origin.

TSTypeID is an integer identifier that indexes a specific type of time series information, as described by the remaining attributes in the TSType table. For a single TSTypeID value in the TSType table, there are zero-to-many records in the TimeSeries table that are described by this time series type.

Variable is a string field that describes what is being measured or calculated. Possible values for Variable include rainfall, streamflow, or total organic carbon concentration.

Units is a string field that describes the units that the measured or calculated data is in. Possible values for Units include cfs. or mg./L.

A useful way of distinguishing types of time series is in terms of the concepts regular and irregular. A regular time series stores data for regularly spaced (uniform interval) time points, while an irregular time series stores data for arbitrary time points (nonuniform intervals). This distinction affects the types of analyses that can be performed on a given data set. Regular time series are appropriate for applications that record entries at predictable time points, such as stream discharge information, which is collected every day. Irregular time series are appropriate when the data arrives unpredictably, such as water-quality samples taken intermittently. Irregular time series can also occur when the duration of a time series changes. For example, during a period of flooding, the frequency of time series measurements for a gaging station may be increased to provide more data for the event. IsRegular is a Boolean field that denotes whether or not the time series records for a particular time series type are spaced at regular or irregular intervals in time. A value of True indicates regular intervals, while a value of False indicates irregular intervals.

The TSDateTime associated with a TimeSeries element marks the beginning of the interval to the next TSDateTime. For regular interval time series data (IsRegular = True), the time interval between these two data points is the TSInterval. The TSInterval field is represented by a coded value domain called TSIntervalType, which provides some of the more commonly used time series intervals (30Minute, 1Day, and so on). The TSInterval is undefined for irregular interval time series data.

Time series data types

Hydrologic processes operate continuously in time, but are measured at particular time points. Depending on the variable being measured, the result may represent the instantaneous value of the variable, such as for streamflow discharge, or it may represent the amount that has accumulated since the last measurement was made, such as for rainfall. In addition, hydrologic data is typically summarized as daily or monthly values even when the data is measured more frequently. These distinctions lead to the definition of the following six types of hydrologic data stored in Arc Hydro as a coded value domain called TSDataType.

1. Instantaneous data—A condition at a given instant of time
2. Cumulative data—The accumulated value since the beginning of the record
3. Incremental data—The difference in cumulative values at the beginning and end of a time interval
4. Average data—The average rate over a time interval, calculated as the incremental value divided by the duration of the data interval

Chapter 7: Time series

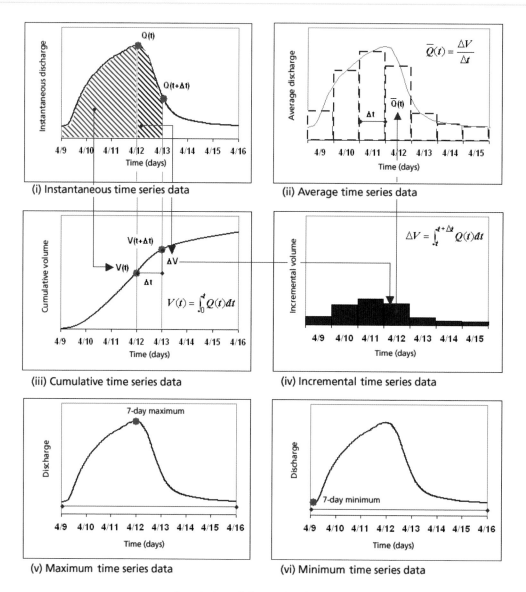

(i) Instantaneous time series data

(ii) Average time series data

(iii) Cumulative time series data

(iv) Incremental time series data

(v) Maximum time series data

(vi) Minimum time series data

Description of time series data types

5. Maximum data—The maximum value of a variable in a time interval

6. Minimum data—The minimum value of a variable in a time interval

These data types can also be illustrated graphically, and defined mathematically. The following example is constructed using one week of 15-minute streamflow data from the Guadalupe River at Victoria, Texas:

Instantaneous time series data: The data represent values that apply for a given instant in time, such as discharge on April 12 at 1:00 A.M.

Average time series data: The data represents the average value over a given period in time, such as average daily streamflow.

Cumulative time series data: The data represents the cumulative value of a variable measured or calculated at a given instant in time, such as cumulative volume on April 12 at 1:00 A.M.

Incremental time series data: The data represents the incremental value of a variable over a given period in time, such as accumulated volume on April 12.

Maximum time series data: The data represents the maximum value over a given period in time, such as seven-day maximum streamflow.

Minimum time series data: The data represents the minimum value over a given period in time, such as seven-day minimum streamflow.

Origin indicates whether the time series values were recorded (measured) or generated (calculated). The Origin field is represented by a coded value domain called TSOrigins, whose possible values are "Recorded" or "Generated." Recorded means that the data was recorded at a monitoring point, while generated means that the data was derived from a hydrologic simulation model of the region.

Using Arc Hydro time series

Several examples of applying time series in Arc Hydro are now presented, including loading data by hand, using a custom tool to automatically load data, and applications of the Time Series data in hydrologic modeling.

Creating an Arc Hydro time series from raw data

Users can create an Arc Hydro time series by using Nexrad radar rainfall data. The following figure shows the Guadalupe basin and the portion of an area for which the Nexrad hourly rainfall data is available.

The Nexrad data is defined on a shapefile of 2,014 polygon cells with hourly rainfall data for each cell for October 13, 2001, when a large flood occurred in the Guadalupe basin. The first step is that the shapefile describing Nexrad data cells is imported into a geodatabase as the Arc Hydro feature class HydroResponseUnit. The Nexrad identifier for this polygon is assigned to the attribute HydroCode, while the Arc Hydro HydroID field of the feature class is used to relationally link each cell with the rainfall time series recorded in that cell as stored in the TimeSeries table.

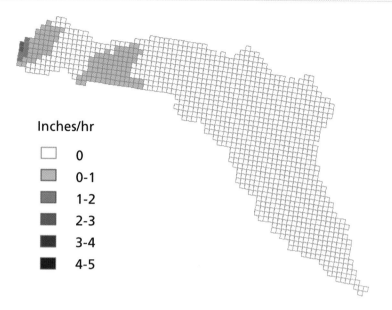

Inches/hr

☐ 0
▨ 0-1
▨ 1-2
▨ 2-3
■ 3-4
■ 4-5

Rainfall 1 A.M. to 2 A.M., October, 13, 2001

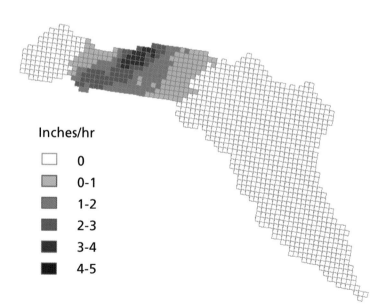

Inches/hr

☐ 0
▨ 0-1
▨ 1-2
▨ 2-3
■ 3-4
■ 4-5

Rainfall 3 A.M. to 4 A.M., October, 13, 2001

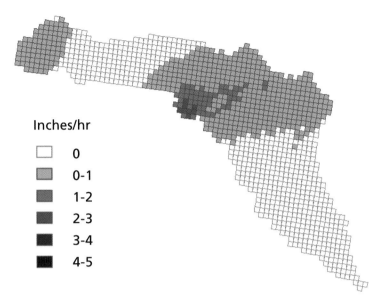

Inches/hr

☐ 0
▨ 0-1
▨ 1-2
▨ 2-3
▨ 3-4
▨ 4-5

Rainfall 5 A.M. to 6 A.M., October, 13, 2001

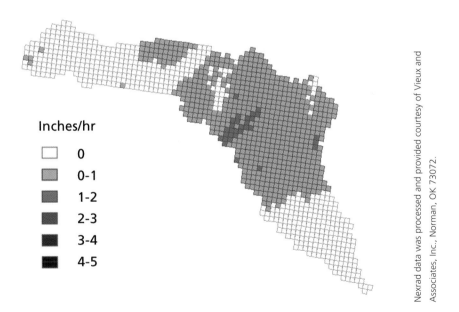

Inches/hr

☐ 0
▨ 0-1
▨ 1-2
▨ 2-3
▨ 3-4
▨ 4-5

Nexrad data was processed and provided courtesy of Vieux and Associates, Inc., Norman, OK 73072.

Rainfall 7 A.M. to 8 A.M., October, 13, 2001
These maps show Nexrad rainfall data for four one-hour periods during the storm of October 13, 2001.
The Arc Hydro time series display tool maps geospatial data that varies through time.

The second step is to create an ArcHydro TimeSeries table using the time series associated with the Nexrad cells. The Nexrad data contains each single date/time value in a separate field, and the Arc Hydro TimeSeries table needs all the date/time information in a single field, TSDateTime. The Arc Hydro time series also requires all the recorded rainfall values in a single column, TSValue. So the user has to modify the input data file using Microsoft Excel or any other spreadsheet program. If Microsoft Excel is used, the number of rows is limited to 65,536 per spreadsheet. In this case there are 2,014 features and each feature has 41 values, resulting in 82,574 records. Several Microsoft Excel sheets can be used to process the data. The result can then be written to a text file and imported into the geodatabase using the Load Data option in ArcCatalog.

The final step is populating the TSType table with information appropriate for this type of data.

TSTypeID = 3 (since types 1 and 2 have already been used for streamflow discharge and gage height)
Variable = Nexrad Rainfall
Units = Inches
IsRegular = True
TSInterval = 1Hour
DataType = Incremental
Origin = Recorded

OBJECTID*	TSTypeID*	Variable	Units	IsRegular	TSInterval	DataType	Origin
1	3	Nexrad Rainfall	Inches	True	1Hour	Incremental	Recorded

Record: 1   Show: All Selected   Records (0 out of 1 Selected.)   Options

TSType table for Nexrad radar rainfall data

Once the data is loaded into Arc Hydro, it is available for any type of analysis or display in ArcGIS, or in a related application such as Microsoft Excel.

Automatically extracting time series

The manual process of data loading just described involves quite a few steps, and for time series data that needs to be imported regularly into Arc Hydro, it is helpful to have a customized tool developed to do that. This section explains how daily streamflow records can be extracted for a number of gaging stations in the Guadalupe basin using a customized application developed in Visual Basic.

The Arc Hydro feature data set for the Guadalupe basin contains all the USGS gaging stations in the MonitoringPoint layer which has a HydroCode field that stores the USGS site number for each point.

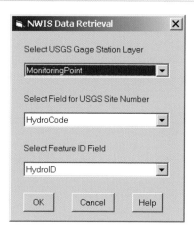

This customized interface is a tool that can be used to extract time series data from the National Water Information System.

Users can specify the time period for the time series record using the Arc Hydro NWIS data tool included on the CD–ROM at the back of this book.

The HydroID field in the MonitoringPoint layer is used to populate the FeatureID field in the TimeSeries table. The TimeSeries table and the feature layer are related to each other through HydroID in the feature layer and FeatureID.

After all the information is entered, the Arc Hydro NWIS data tool extracts the time series data from the USGS National Water Information Web site. The TSType and the TimeSeries table in ArcGIS are populated as shown below. Since this tool downloads only streamflow data, the TSTypeID has only one value, 1 in this case.

Attributes of TSType

OBJECTID*	TSTypeID*	Variable	Units	IsRegular	TSInterval	DataType	Origin
1	1	Daily Streamflow	cfs	True	1Day	Average	Recorded

Record: 1 Show: All Selected Records (0 out of 1 Selected.) Options

TSType table for the streamflow data

Attributes of TimeSeries

OBJECTID*	FeatureID*	TSTypeID*	TSDateTime	TSValue
485537	12000033	1	5/11/2000	671
485538	12000033	1	5/12/2000	581
485539	12000033	1	5/13/2000	603
485540	12000033	1	5/14/2000	586
485541	12000033	1	5/15/2000	766
485542	12000033	1	5/16/2000	967
485543	12000033	1	5/17/2000	750
485544	12000033	1	5/18/2000	614
485545	12000033	1	5/19/2000	596
485546	12000033	1	5/20/2000	893
485547	12000033	1	5/21/2000	1320
485548	12000033	1	5/22/2000	823
485549	12000033	1	5/23/2000	950

Record: 363847 Show: All Selected Records (0 out of 382417 Selected.)

TimeSeries table for the streamflow data

An application example

Once time series data has been loaded into Arc Hydro, it can be used for hydrologic queries and modeling. In a simple GIS application for hydrology, the user can click on a Monitoring-Point feature, perhaps representing stations positioned in and around a catchment of interest or on monitoring points located along a river network, to view and possibly manipulate the time series associated with the feature.

A user can access a tool from a customized time series toolbar, then point to a feature from the HydroPoint feature class. This tool knows about the relationship established between the HydroPoint feature class and the associated TimeSeries class and so knows which time series data to access. The tool can build a graph or table of the data on the fly.

An advanced example

In a more advanced application example, input data is provided to a hydraulic model used for flood analysis. To do the flood analysis, the hydrologist needs to provide the model with the correct discharge in the river. To calculate the discharge, the hydrologist looks at the rainfall data of each subcatchment contributing to the river to determine the total runoff from the rainfall. The hydrologist needs time series data for rainfall, representing the entire catchment. This is done by creating mean area weighted time series based on Thiessen weights for each station within or close to the catchment. The hydrologist can develop a time series software extension to ArcMap for providing the functionality and data access to calculate the mean annual rainfall for each station. From these values a generated mean area weighted time series for all the subcatchments can be generated, then distributed throughout the catchment based upon the weights calculated by the Thiessen method. The final time series is then provided to the modeling software for calculating the total discharge in the river.

To provide the final time series for input to the modeling software, the user creates an ArcMap project and builds a database based on the Arc Hydro data model. All the necessary data is imported into the database, including time series data files. It includes a database of Catchment polygon features, HydroJunctions and MonitoringPoint point features, the HydroEdge (river network) features, and all of the TimeSeries data tables that correspond to the point features in the MonitoringPoint feature class.

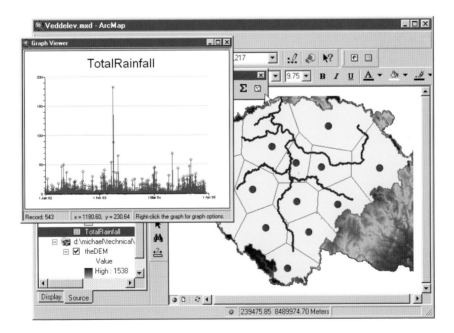

Total rainfall TimeSeries data table is added to ArcMap

From the TimeSeries data tables, the user creates a time series for the given period for each station, in each subcatchment. Note that in this case a separate TimeSeries data table is created for each MonitoringPoint feature by defining a query as [FeatureID = xxxxxxxx], where xxxxxxxx represents the HydroID of each feature.

The application accesses each catchment feature, and from each of the time series data sets within a catchment calculates a mean annual rainfall from all the stations whose rainfall applies to the catchment.

Thiessen polygons can be constructed to determine the weight of each station based upon its distribution over the entire catchment area.

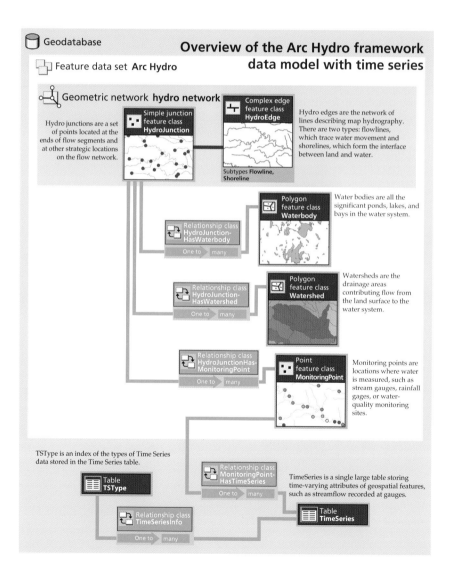

Utilizing these weights and applying them to the mean annual rainfall, a new mean area weighted TimeSeries data table is created for the entire watershed. This information can then be used to calculate the total discharge of a catchment needed by the modeling software.

The Arc Hydro framework UML with time series added to it provides a simple way of constructing such applications.

Data dictionary

The diagram on the following page summarizes the object and feature classes in the time series component of Arc Hydro, and their interrelationships. All the classes shown are available for loading data because they have inherited all the attributes from classes located above them in the UML hierarchy. The attributes shaded in blue are ESRI standard attributes, while those shaded in white are Arc Hydro attributes. The terms used here are defined in the glossary at the back of this book.

Coded value domain
TSOrigins

Description
  Field type *Integer*
  Split policy *Default Value*
Merge policy *Default Value*

Code	Description
1	Recorded
2	Generated

Coded value domain
Boolean

Description
  Field type *Integer*
  Split policy *Default Value*
Merge policy *Default Value*

Code	Description
1	True
0	False

Coded value domain
TSDataType

Description
  Field type *Integer*
  Split policy *Default Value*
Merge policy *Default Value*

Code	Description
1	Instaneous
2	Cumulative
3	Incremental
4	Average
5	Maximum
6	Minimum

Coded value domain
TSIntervalType

Description
  Field type *Integer*
  Split policy *Default Value*
Merge policy *Default Value*

Code	Description
1	1Minute
2	2Minute
3	3Minute
4	4Minute
5	5Minute
6	10Minute
7	15Minute
8	20Minute
9	30Minute
10	1Hour
11	2Hour
12	3Hour
13	4Hour
14	6Hour
15	8Hour
16	12Hour
17	1Day
18	1Week
19	1Month
20	1Year
99	Other

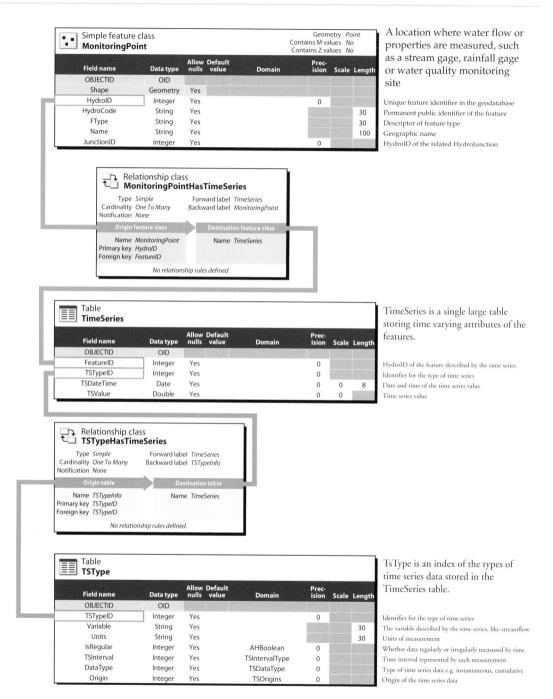

Simple feature class
MonitoringPoint

Geometry *Point*
Contains M values *No*
Contains Z values *No*

A location where water flow or properties are measured, such as a stream gage, rainfall gage or water quality monitoring site

Field name	Data type	Allow nulls	Default value	Domain	Precision	Scale	Length	
OBJECTID	OID							
Shape	Geometry	Yes						
HydroID	Integer	Yes			0			Unique feature identifier in the geodatabase
HydroCode	String	Yes					30	Permanent public identifier of the feature
FType	String	Yes					30	Descriptor of feature type
Name	String	Yes					100	Geographic name
JunctionID	Integer	Yes			0			HydroID of the related HydroJunction

Relationship class
MonitoringPointHasTimeSeries

Type *Simple*
Cardinality *One To Many*
Notification *None*

Forward label *TimeSeries*
Backward label *MonitoringPoint*

Origin feature class	Destination feature class
Name *MonitoringPoint*	Name *TimeSeries*
Primary key *HydroID*	
Foreign key *FeatureID*	

No relationship rules defined

Table
TimeSeries

TimeSeries is a single large table storing time varying attributes of the features.

Field name	Data type	Allow nulls	Default value	Domain	Precision	Scale	Length	
OBJECTID	OID							
FeatureID	Integer	Yes			0			HydroID of the feature described by the time series
TSTypeID	Integer	Yes			0			Identifier for the type of time series
TSDateTime	Date	Yes			0	0	8	Date and time of the time series value
TSValue	Double	Yes			0	0		Time series value

Relationship class
TSTypeHasTimeSeries

Type *Simple*
Cardinality *One To Many*
Notification *None*

Forward label *TimeSeries*
Backward label *TSTypeInfo*

Origin table	Destination table
Name *TSTypeInfo*	Name *TimeSeries*
Primary key *TSTypeID*	
Foreign key *TSTypeID*	

No relationship rules defined.

Table
TSType

TsType is an index of the types of time series data stored in the TimeSeries table.

Field name	Data type	Allow nulls	Default value	Domain	Precision	Scale	Length	
OBJECTID	OID							
TSTypeID	Integer	Yes			0			Identifier for the type of time series
Variable	String	Yes					30	The variable described by the time series, like streamflow
Units	String	Yes					30	Units of measurement
IsRegular	Integer	Yes		AHBoolean	0			Whether data regularly or irregularly measured by time
TSInterval	Integer	Yes		TSIntervalType	0			Time interval represented by each measurement
DataType	Integer	Yes		TSDataType	0			Type of time series data e.g. instantaneous, cumulative
Origin	Integer	Yes		TSOrigins	0			Origin of the time series data

## chapter eight

Hydrologic modeling

Venkatesh Merwade, University of Texas at Austin
David Maidment, University of Texas at Austin

Integrating GIS and hydrologic modeling involves connecting geospatial data describing the physical environment with hydrologic process models describing how water moves through the environment. Observed time series of water resources data at monitoring points are needed to verify hydrologic models. Arc Hydro provides a robust environment for integrating geospatial and time series data for water resources with simulation models for water flow and quality. A variety of interfaces exist for accessing Arc Hydro data, including ArcGIS, Microsoft Access, Microsoft Excel, and direct interface programming with Visual Basic. Dynamic linked libraries provide a secure, efficient way to link process models with Arc Hydro data.

Integrating GIS with hydrologic modeling

Water is constantly flowing through the landscape in response to weather events—at times a deluge during a storm, at other times a trickle during a drought. Most often, water flows in the regular patterns that we come to expect from rivers and streams. Water flow carries microorganisms, dissolved chemicals, and sediment that define water quality. We need to understand these flow and quality patterns, and how they change under different management plans for water resources. Hydrologic simulation models provide a representation of water flow and quality for rivers, streams, lakes, bays, and estuaries. GIS supports hydrologic analysis and modeling by describing the physical environment through which water flows.

A hydrologic simulation model may be formally coded in a programming language or it may be developed in a spreadsheet format. Spreadsheets are particularly useful for analysis of observed hydrologic data and for developing prototype simulation models used in a local setting. For a long time, hydrologists have wanted to apply GIS data in a straightforward manner using spreadsheets. Arc Hydro facilitates this goal since the Arc Hydro personal geodatabase is a Microsoft Access file, and there is a link between Microsoft Excel and Microsoft Access allowing Arc Hydro data to be directly viewed and manipulated in Microsoft Excel.

Formally coded hydrologic simulation models describe particular hydrologic processes and are intended to be applied at any geographic location. Some of the most famous of these models were developed during the 1960s and have been in use for decades, such as the HEC-1 and HEC-2 flood simulation models from the Hydrologic Engineering Center of the U.S. Army Corps of Engineers. As new software engineering approaches have emerged, they have been adopted for hydrologic and hydraulic modeling, such as the HEC-HMS and HEC-RAS simulation models. These are written in object-oriented programming languages and are the successor programs to HEC-1 and HEC-2. During the 1990s, special-purpose GIS interfaces were constructed to supply geospatial data for popular hydrologic models such as the HEC-GeoHMS and HEC-GeoRAS systems, which were developed using the Avenue language in ArcView 3.

GIS can be used in various ways to support hydrologic modeling. GIS can:

- Manage data—GIS performs basic geospatial data-management tasks (data storage, manipulation, preparation, and extraction) and spatial data processing (overlays and buffering).
- Extract parameters—GIS provides characteristic properties of watersheds and river reaches for hydrologic modeling.
- Provide visualization—GIS displays data, either before the hydrologic analysis is performed to verify the basic information, or after the analysis to evaluate the results. For example, floodplain mapping in GIS shows the extent of areas damaged by floods .
- Model surfaces—GIS delineates watersheds and represents channel shapes based on digital terrain or elevation models.
- Develop interfaces—Map-based interfaces to hydrologic models can be developed using GIS tools.

The EPA Basins program for water-quality simulation is a particularly sophisticated integration of GIS and hydrologic modeling, built using ArcView 3. Basins comes with a comprehensive set of structured data covering the whole United States using eight-digit Hydrologic Unit

Code watersheds as data-packaging units. The Basins program contains tools for data synthesis and assessment, as well as preparing input data for several water-quality simulation models.

The applications just described involve the GIS and hydrologic model operating independently of one another and interacting through text files. This well-tested process will continue to be implemented within ArcGIS as new application programs are developed in Visual Basic to replace ones previously created in ARC Macro Language for workstation ARC/INFO or in Avenue for ArcView 3. Arc Hydro significantly facilitates this process because many basic functions such as watershed and stream network delineation are already implemented in the Arc Hydro toolset, and the core Arc Hydro data structure formulates GIS data in an appropriate format for water resources analysis.

Moreover, the advent of ArcGIS and Arc Hydro offer some new possibilities for integration of GIS and hydrologic modeling, which are explored in this chapter. Since Arc Hydro is relatively new, a limited amount of knowledge and experience is currently available on this subject, and the discussion presented in this chapter is relatively brief. Additional information about how to implement the procedures described here is contained on the Arc Hydro CD–ROM.

Hydrologic information systems

Traditionally, the term "water resources data" has meant the time series of water measurements recorded at gages and monitoring sites. As GIS has become more popular in water resources, geospatial information for water resources has become available, including GIS data layers for stream networks, watersheds, water bodies, and measurement station locations. Arc Hydro combines geospatial and temporal data in a relational database format.

Hydrologic information system

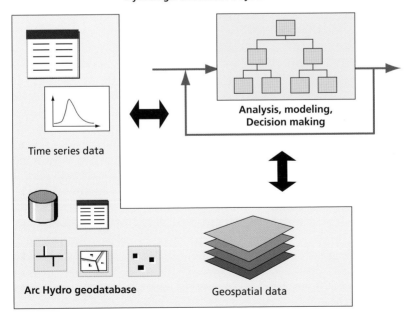

Time series data

Analysis, modeling, Decision making

Arc Hydro geodatabase

Geospatial data

What is emerging is a new kind of technology that might be called a "hydrologic information system," defined as a structured database of geospatial and temporal water resources data, combined with tools for information processing, which supports hydrologic analysis, modeling, and decision making. Analysis and modeling may be performed in several ways: by using ArcGIS, by using spreadsheet programs such as Microsoft Excel, by using callable routines in a dynamic linked library (DLL), or by using independent hydrologic models.

What distinguishes a hydrologic information system from earlier approaches is that supporting data is contained in a formally structured relational database, and that geospatial and temporal information is integrated. The resulting structure can support several hydrologic simulation models operating cohesively, rather than having each model operating separately.

Interfaces for hydrologic modeling

ArcGIS was designed to move GIS more into the mainstream of information technology. Instead of using a customized data file structure, ArcGIS data is stored in a standard relational database. Instead of having a special GIS programming language, ArcGIS uses Visual Basic as the programming language for its interface, similar to the Microsoft Office application programs. The result of these design choices for ArcGIS is that Arc Hydro data stored in a geodatabase is directly accessible by other application programs without even going through ArcGIS.

A hydrologic information system can be accessed using at least four different interfaces: ArcGIS, Microsoft Access, Microsoft Excel, and by programming Visual Basic. These interfaces are windows on the Arc Hydro data and hydrologic analysis programs. Each interface has its own characteristics, advantages, and limitations.

In a Visual Basic interface case, an independent Visual Basic application interacts with the Arc Hydro data and with the hydrologic model to perform hydrologic analysis. Visual Basic is capable of directly reading Microsoft Access files. This approach is particularly appropriate when the focus is on hydrologic modeling and Arc Hydro is only needed as a source of data to support the model operation. However, there is no map interface to the model.

In a Microsoft Excel interface case, Arc Hydro data is directly imported from the Arc Hydro geodatabase in Microsoft Access to Microsoft Excel. Hydrologic analysis is carried out with the normal functionality of Microsoft Excel. If more sophisticated applications are required, they can be created using Visual Basic for Applications for Microsoft Excel, possibly calling routines in a dynamic linked library. An example of using a dynamic linked library from Microsoft Excel is presented later in this chapter. Because many hydrologists are familiar with the Microsoft Excel interface, they do not need to do custom programming.

Visual Basic for Applications for Microsoft Access can be used to interact with the Arc Hydro geodatabase to perform hydrologic analysis. This approach can do more than a comparable simulation program in Microsoft Excel because Microsoft Access maintains a tighter control over changing and manipulating the data than does Microsoft Excel. Microsoft Excel is very good for developing prototypes, but production programs operate more efficiently on Microsoft Access databases. Users should be cautioned that the geodatabase is carefully structured to operate in

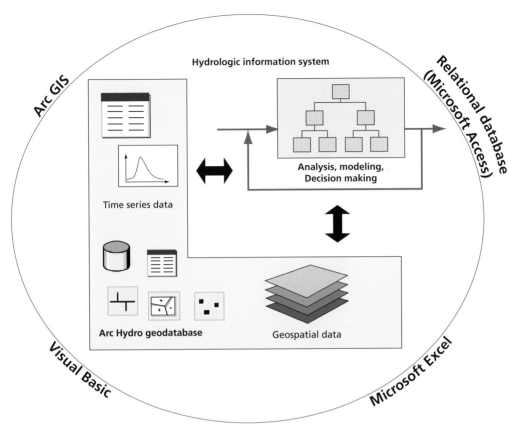

Interfaces for hydrologic modeling

ArcMap. So while information can be read from the geodatabase in Microsoft Access, it is advisable to use ArcGIS utilities to add information back into the geodatabase.

ArcGIS can be customized using Visual Basic. Hydrologists can create custom applications in ArcGIS as built-in functions that use Arc Hydro data to perform a hydrologic analysis or run a hydrologic model. This is the most sophisticated approach and has the advantage of using the ArcMap interface, so that results can be visualized in the map environment. However, programming the map interface in ArcGIS is more complicated than just manipulating Arc Hydro data tables in Microsoft Access or Microsoft Excel.

Remarkably, at one level, it does not even matter which of these interfaces is used because the computer code that represents the functioning of hydrologic processes operates the same in all of these environments. What differs among them is the way the data is put into and taken out of the program. Core Arc Hydro data is available to the hydrologist through a variety of interfaces, and depending on the nature of the application, one or more of these interfaces may be used to analyze it.

Hydrologic modeling using Arc Hydro data

Hydrologic analysis and modeling involves solving the mathematical equations that describe water flow and quality. This can be done in three ways:

- Intrinsic modeling—Using the internal functionality of a particular application for hydrologic analysis. For example, within ArcGIS map algebra or attribute calculations can be used to calculate runoff maps from rainfall maps. Or, Microsoft Excel can be used to write a simple hydrologic simulation model.
- Dynamic linked library (DLL)—The hydrologic process representation is contained in a separate code module, or DLL file, that can be written in Visual Basic or in another programming language such as C or FORTRAN. This DLL file can be called from the user interface using Visual Basic. DLL files are used to customize ArcGIS.
- Independent modeling—The hydrology model operates as an independent executable program, and Arc Hydro data is input to the program and received from it using structured text files. To date, this is the most common way that GIS interfaces have been constructed for hydrologic models.

In each case, the geospatial and temporal water resources data stored in Arc Hydro provides the information needed to support the functioning of the hydrologic model.

Intrinsic hydrologic modeling

Hydrologic modeling using an intrinsic model involves using the internal functionality of an application, such as ArcGIS or Microsoft Excel, to perform hydrologic simulation. This type of modeling does not involve an interface design or programming effort. However, the user needs to know the functionality available with the application. For example, an intrinsic ArcGIS model can be constructed for calculating the mean annual pollutant load from a watershed. Data about precipitation values, land use, and pollutant concentrations is required. The methodology for determining the pollutant load is described by Quenzer and Maidment (1998).

First, the runoff is calculated using a precipitation grid and land-use data. This can be done using an empirical rainfall-runoff function, whose parameters are defined for different land uses. For example, urban land has more runoff than agricultural or forest land. Similarly, expected pollutant concentration values are defined for each land use. The runoff values are multiplied by the concentration values to get pollutant load from each DEM cell or land-use polygon and summed using an accumulation function to get the total pollutant load from the watershed. This process is well-suited to raster calculation in GIS, but when the size of the raster is large, the volume of data to be processed can become overwhelming. Arc Hydro provides another way to approach this problem since the Arc Hydro catchments provide a fine-scale tessellation of the landscape, and Arc Hydro software tools have been developed that mimic the accumulation functions present in the raster GIS environment. The user can employ the ArcGIS field calculator to determine runoff and pollutant load in each catchment, then accumulate the total loads going downstream using the Arc Hydro Accumulate tool described in chapter 4 and available on the CD-ROM.

One of the intriguing and largely unexplored avenues for intrinsic modeling of hydrologic processes within ArcGIS is to develop customized behaviors for Arc Hydro features. This is done using C++ code attached directly to the features at the time that the Arc Hydro schema is applied to data in the geodatabase. For example, a customized watershed feature may carry a function that simulates its behavior when rainfall occurs. This function could read the rainfall data from the Arc Hydro TimeSeries table, determine the corresponding runoff, and write the result back to the TimeSeries table. In this way, an Arc Hydro model is not simply a data model as it is now, but also has intrinsic simulation capabilities. It is possible that powerful new map-based hydrologic simulation programs could be developed using custom Arc Hydro features.

Dynamic linked library

A dynamic linked library, or DLL, is a library of executable functions or modules that can be used by other programs. These functions can be written in languages other than Visual Basic. The user invokes a particular subroutine within the DLL by passing the required type and number of arguments to the DLL. Using a DLL, information can be passed back and forth between programs as a set of parameters. For example, a DLL can be written that calculates a water-surface profile. In this case the DLL takes water elevation (stage) data at the monitoring stations of a river reach as the input from GIS, does the flow routing calculation, and returns the water-surface profile for the entire river reach, which can then be displayed in GIS. Application integration through DLLs is tighter and faster than running an external executable program. In the context of Arc Hydro, DLL functionality allows the user to do simple as well as complicated hydrologic calculations through an interface.

Some time ago, the Hydrologic Engineering Center of the U.S. Army Corps of Engineers took its HEC-1 flood simulation model, isolated from the general code the FORTRAN sub-routines that perform particular functions, and documented the result as a program library called

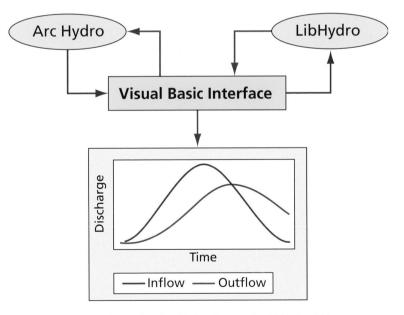

River routing using Arc Hydro data and a LibHydro DLL

LibHydro (Hydrologic Engineering Center 1995). LibHydro contains subroutines in the following categories: precipitation, loss rates, unit hydrographs, base flow, routing, utility, error handling, and SI/English unit conversions. The subroutines are written in FORTRAN and most of them conform to the FORTRAN 77 standard, with selected extensions that conform to the FORTRAN 90 Standard.

It is possible to transform this LibHydro FORTRAN subroutine library into a DLL and to call individual subroutines in the DLL from Visual Basic. The interface protocol of the DLL translates the definition of the parameters passed from the Visual Basic calling program into an appropriate definition for the FORTRAN subroutine, executes the FORTRAN subroutine within the DLL, then translates the output results into an appropriate format for the Visual Basic calling program. These translations happen in memory rather than with text files, so execution is rapid.

The Muskingum method is a hydrograph routing method used for channel flow routing and is one of the subroutines available with LibHydro DLL. The input data required by the Muskingum method is the inflow hydrograph at the upstream end of a river reach, Muskingum routing parameters K and X, and the number of subreaches into which the river will be divided for analysis. In this case, the Visual Basic interface reads Arc Hydro data, calls the LibHydro DLL, passing it the inflow hydrograph and routing parameters, and receives the outflow hydrograph as an output. This hydrograph can be stored in Arc Hydro if desired. The Visual Basic interface to Arc Hydro data and the LibHydro DLL can be operated from ArcGIS, Microsoft Excel, Microsoft Access, or from an independent Visual Basic program, as described earlier.

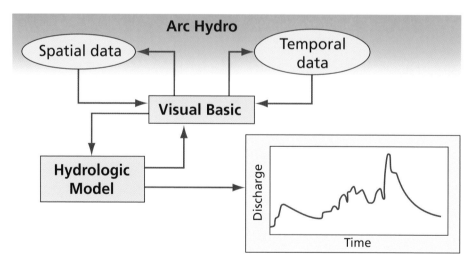

Independent hydrologic model with a Visual Basic interface to Arc Hydro

Independent hydrologic model

An independent hydrologic model is a set of coupled hydrologic processes that simulates a whole phenomenon. Instead of a single function such as routing water through a river reach, an independent model contains a set of linked functions that translate rainfall into runoff over a set of watersheds, and route the flow downstream through the river and stream network. The model runs externally from the GIS, and has its own user interface. A Visual Basic program can be written to take geospatial and temporal water resources data from Arc Hydro and write them into text files suitable for reading by the independent model. The model is executed and produces text files as output. These may be read back into the Arc Hydro geodatabase for display purposes if necessary. Furthermore, if the hydrologic model itself is written using a Visual Basic interface, then that interface can read and write information directly to the Arc Hydro geodatabase without going through text file translation.

Hydrologists want to be able to use GIS data both to operate traditional, well-proven models, and also to develop new, more effective analyses and models. The approaches to hydrologic modeling described above provide a variety of methods by which these goals can be accomplished using Arc Hydro data. Since ArcGIS uses a standard interface language, Visual Basic, and stores its data in a standard relational database, access to the data is more direct than was the case with earlier versions of ArcInfo and ArcView. It is no longer necessary to connect hydrologic models with GIS data through text files, as direct data sharing is possible among a set of applications or interfaces, which can all view the Arc Hydro geodatabase and invoke hydrologic analysis routines stored as dynamic linked libraries.

It should be noted that the discussion in this chapter applies to the personal geodatabase, where Arc Hydro data is stored as a Microsoft Access file. The larger enterprise geodatabase applications of ArcGIS or Arc Hydro operate on fully configured relational databases such as Oracle or Informix®, with which data exchange is more complex. In ArcGIS, this data exchange is accomplished with the Spatial Database Engine™ (SDE®). Also, Visual Basic is not the only language that can be used to build interfaces to Arc Hydro data. This can also be done using Visual C++® and other COM-compliant languages.

As to how the linking of GIS and hydrologic modeling will evolve, only future years will tell the story. What is clear, however, is that the technology of ArcGIS and its customization for water resources in Arc Hydro offer to hydrologists a stronger, more accessible way to manipulate GIS data, to synthesize that with water resources time series data, and to tightly link the data with hydrologic analysis and modeling routines. These new capabilities open up many exciting opportunities for more effective analysis of water resources systems.

Implementing Arc Hydro

Steve Grise, ESRI
David Arctur, ESRI
Bob Booth, ESRI

Arc Hydro provides a set of objects and features that can be used as a starting point for many water resources projects. Arc Hydro can be implemented with no modifications, or it can be highly customized to fit your specific needs. This chapter describes how to create Arc Hydro data sets for typical hydrologic projects. For enterprise data management systems, database administrators can use Computer Aided Software Engineering (CASE) tools to create custom objects and design a new UML diagram to generate a geodatabase schema.

A simple Arc Hydro implementation

Arc Hydro comes in several variants to simplify application. The simplest, the Arc Hydro framework, has five feature classes to represent the hydro network, watersheds, water bodies, and monitoring points. If you want to add time series, you can select the Arc Hydro framework with the time series option, which enables TimeSeries tables to be stored for monitoring points. It is probably best to begin your work with Arc Hydro with one of these simple designs so that

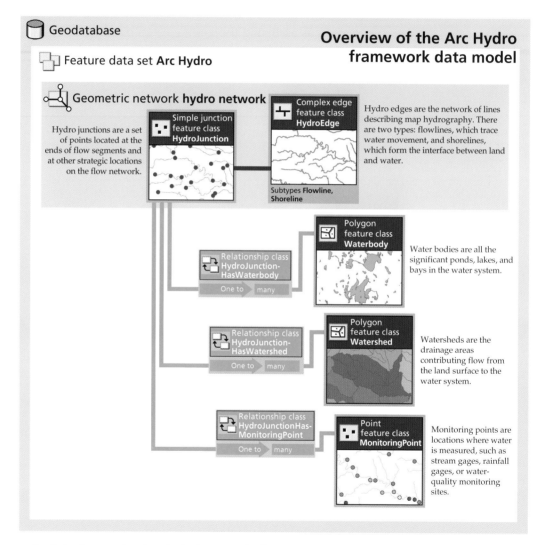

Analysis diagram showing the basic Arc Hydro framework. Database designers can add classes and attributes to this framework.

you become familiar with the process of building the geodatabase, applying the schema to it, and developing values for the Arc Hydro attributes. The CD–ROM included with this book contains a detailed set of instructions to help you accomplish these tasks. Once you've mastered the process of applying Arc Hydro to a small data set, it will be easier to implement additional Arc Hydro components.

To help you implement Arc Hydro, a customized Arc Hydro toolset for ArcGIS has been developed and the current version is contained in the CD–ROM, along with a tutorial on how to use the tools and a sample data set from the Guadalupe basin on which you can apply the tools. These tools are particularly helpful for raster analysis of DEMs to produce Arc Hydro drainage features, and to do such tasks as assigning the HydroID and building relationships between HydroJunctions and other feature classes. The toolset also has some analysis functions that enable you to do elementary hydrologic calculations and trace hydrologic features in the landscape both upstream and downstream. The Arc Hydro toolset will continue to be developed and later versions of the tools will be available on the Web site of the GIS in Water Resources Consortium at www.crwr.utexas.edu/giswr.

This chapter presents a summary of the process of implementing Arc Hydro, and also of customizing Arc Hydro by adding or deleting classes and attributes so that the data structure meets your needs. In many cases, the content of the Arc Hydro data model is sufficient for implementation in its present form. You can create a geodatabase from a Microsoft Repository database and use Arc Hydro without doing additional customization, as described later in this chapter.

General implementation procedure

To implement the data model from the Microsoft Repository containing Arc Hydro, the following steps are required. More detail on steps 3–8 can be found in the ESRI book *Building a Geodatabase* (MacDonald 2001*)*. These steps assume you are familiar with ArcGIS and that you are applying the full Arc Hydro data model.

1. Install ArcInfo/ArcEditor™ Desktop, and any analysis extensions you plan to use.
2. Create a personal geodatabase or enterprise geodatabase in ArcSDE™.
3. Create feature data sets with the correct spatial references for your geographic area. The names of the data sets are Hydrography, Drainage, Network, Channel, and Time Series. By default you will get an unknown spatial reference for your data sets, and you cannot change this once the schema is defined.
4. Use the ArcCatalog Schema Wizard to create a physical schema from the Arc Hydro repository database.
5. Use ArcCatalog to edit the database design to match your conceptual design/ project needs.
   a) Delete the classes and attributes that you do not need for this particular application.
   b) Add additional classes and attributes to the basic model.
   c) Add additional information (object and feature classes) that is not directly part of the model but is required for your project. Examples include background layers like county or land-parcel boundaries.
6. Load data into the geodatabase.
7. Deploy the geodatabase. In the case of an enterprise geodatabase, this involves versioning the database and setting up security (if required).

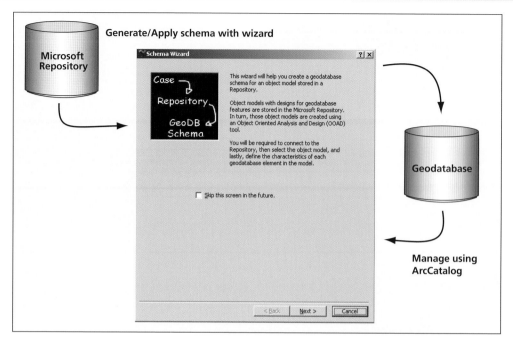

You can create your Arc Hydro database using the Schema Wizard.

Deploying the geodatabase

Once you've built the geodatabase, you will usually need to make it available to people in your organization who use the data. This is true in enterprise projects where multiuser editing is required, and also in water resource management agencies where many project teams will use a common regional database. These people may work with a geodatabase in different ways. Some on your project team may create and edit alternative versions of the database during the design process, analysts may model flows or trace connected parts of the network, and managers may want to view progress on specific project tasks. You can give people access to the information they need, with the tools they need, through ArcCatalog and ArcMap.

Creating Arc Hydro data sets on a typical water resources project

Follow these steps when implementing Arc Hydro for a typical hydrologic study.

Step 1. Define the spatial extent of the study region

For a detailed study, hydrologists usually focus on a specific geographic area. Understanding the extent and available data for your study is the first step in a project.

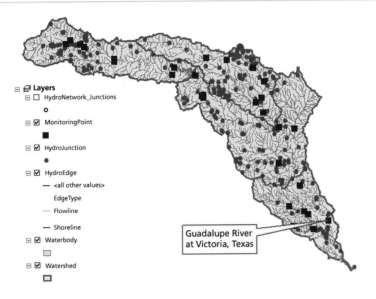

Arc Hydro framework data set for the Guadalupe River basin

Step 2. Define and tailor the Arc Hydro database

To create a database, perform analysis, design, and construction.

To do analysis, complete an assessment of your needs and look at the data you require to complete your project. Understand the map products you will produce and what data elements are required to do analysis.

To design a database, define the components required to adequately model your system. Most people create an analysis diagram for their project as a way to explain the data they use in their project. On some projects this is not necessary, especially if you are using Arc Hydro with few modifications.

The physical database model defines the database schema, class structure of objects, and how rules and relationships are implemented. In most cases, you should build the data model using the Schema Wizard and other tools in ArcCatalog. The wizard reads the structure of the database design from a Microsoft Repository file. Depending on whether you are building a personal geodatabase or an enterprise database on ArcSDE, the wizard then creates and applies database creation statements appropriate to the specific database platform you are working with. While the process sounds complicated, it is actually quite simple to create a database using the Schema Wizard.

You can then use ArcCatalog to add, remove, and modify the database to your project needs. In this type of project we assume you are using a personal geodatabase. For an enterprise database in ArcSDE the process is quite similar, but there are additional things you need to understand while creating the database and loading data. Refer to ArcOnline.esri.com in the White Papers section for additional information on Building Multi-User Geographical Information Systems with ArcInfo 8.

Step 3. Assemble hydrography for the study region
Obtain the basic feature data for your study area by digitizing features from existing maps or converting tabular data inventories into point feature classes. Load the data into the target geodatabase. These are the hydrography feature classes for the project in the Arc Hydro model.

Step 4. Build a hydro network
Load data from the map hydrography and other data sources into the target schema using the simple data loader in ArcCatalog. Recreate the geometric network in ArcCatalog. If you have only a small amount of data, you don't need to go through the process of deleting the network—just use the Object Loader in ArcMap instead. Detailed instructions about this step are contained on the CD–ROM.

Step 5. Define DrainageAreas and attach them to HydroJunctions
The link between the landscape or watershed polygons and the hydro network is through HydroJunctions. HydroJunctions connect edges in the geometric network. The drainage areas, typically watersheds, need to have the correct outlet junction associated with them. This is the way that flow is transferred from the landscape onto the network. You can do this manually or use a tool to populate the JunctionID for each DrainageArea.

Step 6. Attach TimeSeries for key MonitoringPoints
If you are using time series data, now is the time to associate the data with the correct monitoring points. MonitoringPoints are in the Hydrography package. TimeSeries records have an attribute FeatureID that can be associated with the MonitoringPoint HydroID value or a project-specific identifier. You can add additional ID fields to hydro features and the time series objects as required.

Step 7. Define Channel shape
If you need to create Channel shapes for your area of interest, create or load this data. The data typically consists of Profile and CrossSection representations and is often passed on to hydraulic analysis programs for more specific analysis. Managing this data in ArcGIS is a good way to load and validate the data you are using in the context of the rest of your project database. It is also a good way to make sure the data from the project is available for other people to use in the future.

Step 8. Deploy the database and map documents
At this stage you make the data in your geodatabase available for use. ArcCatalog and ArcMap are the two main applications you and others in your organization will use to work with the geodatabase. ArcCatalog lets you manage the database, publish layers with symbology throughout your organization, load data, and create versions of the geodatabase. You make the data in the geodatabase available by placing maps and layer files, which reference the data in the database, in shared folders for your system's various types of users. You can control access to data by creating password-protected connections to the database.

ArcMap allows you to edit data while maintaining network connectivity, trace through the network with a variety of tools, and create maps tailored to specific purposes.

Step 9. Expand the geodatabase with additional data
Often projects will begin with hydrography and gradually more data will become available through the lifetime of the project. Once you are familiar with ArcGIS, whether with the CASE tools or without, you can easily add additional classes, attributes, and load data into your database.

Often people are very concerned at the beginning of their project that the database design be perfect, only to have their project scope and data requirements change over time because of changes in priorities or budgets. The key point to remember in designing and building the project database is that it is easy to modify and enhance the schema over time. The software has been designed to support multiple iterations of design and deployment, so that you can easily implement your project and adapt to future needs.

Using CASE tools and the Unified Modeling Language (UML)

Much of the hard work of analyzing, designing, and building a good Arc Hydro geodatabase has been done for you. But some organizations will need to go through the full analysis and design cycle to build enterprise-level data management systems. Some of these project teams will want to use UML to define and manage their database design.

CASE (Computer Aided Software Engineering) tools and techniques automate the process of developing software and database designs. You can use CASE tools to create new custom objects and to generate a geodatabase schema from a UML diagram. In the CASE tool environment you can create a schema quickly using automated tools. CASE tools provide a way to systematically capture the refinement of your design over time, so you don't need to have dozens of design notes scattered all over your desk.

Two general strategies exist for using UML and CASE tools to design and create a geodatabase. The first involves using UML to define the schema for the geodatabase, generating that schema, then populating the schema with data. The second takes a different approach and

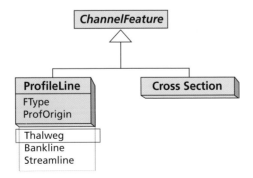

Modeling Thalweg and other types of ProfileLines

involves creating the schema by importing existing data into your geodatabase, building geometric networks, then using CASE tools to apply your UML model to the existing data.

You can also use a combination of the two strategies. Once a schema has been created, you can modify it by modifying the UML model, then reapply the model to the geodatabase schema using the Schema Wizard.

You can work in ArcCatalog before and after running the Schema Wizard to ensure the changes you want to make are correctly applied in the database. Alternatively, you can just use the schema management tools in ArcCatalog to modify the geodatabase schema and ignore UML after the initial schema creation.

Example: Modeling a ProfileLine

There are many different methods to model real-world objects. The following example shows the steps needed to model a common channel feature, a ProfileLine, using Arc Hydro.

ProfileLine is a general class designed to represent the longitudinal profile of a river. If necessary, specific profile lines can be defined: (1) the center or thalweg of the channel, (2) the left and right banklines where the water meets the shoreline, and (3) a streamline that represents any other cartographic line, such as a floodline on the left and right overbank flow areas that represents the general pathway of water flow in the floodplain.

These types of ProfileLines can be implemented as a coded value domain or as a subtype. A coded value domain is a simple list of codes like 1, 2, 3 and a descriptive label.

Another option for ProfileLine type is to make geodatabase subtypes for each of the types. Subtypes provide additional abilities to specify different default values and domains for each subtype of the class. In the case of ProfileLine, you can create subtypes if you have different sources for the attribute "ProfOrigin," the origin of the profile data. You can associate a different list of values for Thalwegs, Banklines, and Streamlines.

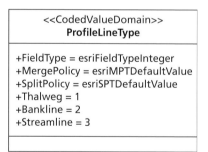

In UML, a coded value domain for ProfileLine

Implementing Arc Hydro using CASE tools

In the previous sections we discussed the process of defining the Arc Hydro data model and typical implementation approaches. This section describes how to implement the Arc Hydro geodatabase using UML and the CASE tool environment. You may use some or all of the described methods, depending on your requirements. The ESRI books *Modeling our World* (Zeiler 1999) and *Building a Geodatabase* (MacDonald 2001) provide directions for designing and implementing custom geodatabases.

Implementing a system using the Arc Hydro geodatabase UML provides the most control over the database content. CASE tools are valuable when multiple projects require a similar database design and are useful for large projects that involve data management in an enterprise environment. This approach requires additional software, Visio® Enterprise, to manage the .vsd files included with this data model on the CD–ROM.

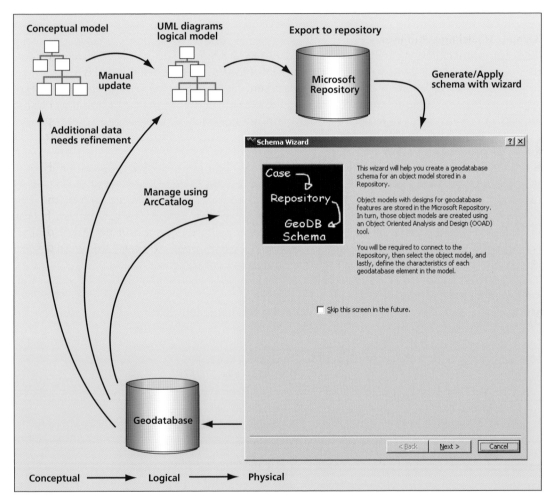

Using CASE tools to design and refine the Arc Hydro data model

Establish a data model

To begin, install the ArcInfo or ArcEditor Desktop and the sample geodatabase design for the Arc Hydro data model from the CD. You also need to install Visio Enterprise.

After you have completed your conceptual design, you can modify the UML to reflect your logical design. From Visio, you can export a Microsoft Repository database. Use ArcCatalog to import the schema from the Microsoft Repository into your geodatabase.

The Arc Hydro data model UML is a diagram that captures a design for a geodatabase. The design itself can be stored in a DBMS (either Microsoft Access or SQL Server) as a Microsoft Repository, which then can be read by ArcCatalog to create a schema for the geodatabase. The repository contains a generalized representation of all the objects (tables or feature classes) showing their inheritance relationships as well as subtypes, domains, default values, relationships, and connectivity rules.

ArcCatalog contains tools to read the Microsoft Repository. The Schema Wizard guides you through the process of creating new feature classes, tables, and other components of the geodatabase. The geodatabase schema can be read directly from the repository. Once the wizard is finished, you will have a schema that you can modify in ArcCatalog as needed.

Refine the geodatabase using ArcCatalog

You can use ArcCatalog to continue defining the geodatabase by establishing how objects in the database relate to one another. This is the simplest and most direct method of implementing Arc Hydro.

Use ArcCatalog to establish relationships between objects in different object classes, add attributes, and associate them with domains. You can continue to use the geodatabase management tools in ArcCatalog to refine or extend a mature database throughout its life. You may want to use a combination of ArcCatalog and UML to maintain your database design.

Implementing Arc Hydro using UML

To implement a data model from Arc Hydro using UML, the following steps are required:

1. Install ArcInfo Desktop and any analysis extensions you plan to use.
2. Create a logical data model in UML.
3. Create a geodatabase and feature data set(s) with the correct spatial reference(s).
4. Generate the physical database model from the logical UML model.
5. Use ArcCatalog or UML or both to edit the schema. Whenever you modify the UML, you need to "reapply the schema" to the existing database.
6. Load the data into the geodatabase. For larger data sets, drop the relationships and networks, use the simple data loader in ArcCatalog, rebuild the network, then reapply the schema.
7. Deploy the geodatabase.

For more specific information on performing these steps, refer to the ESRI books *Using ArcCatalog* (Vienneau 2001) and *Building a Geodatabase* (MacDonald 2001).

Ongoing refinement

As you progress with your projects, you can continue to update your design in UML. For instance, you might generate many project databases from the one design, then discover that you are missing some important attributes. To apply the changes to all of the existing databases, re-apply the schema to each of the databases to add the new fields or make other modifications.

Keep in mind that the process of applying the schema always "adds" to the database. For instance, if you rename a field in the UML model, the Schema Wizard will prompt you whether you want to create a new field or map this field to an existing field. If you add a new field, you will probably need to update that field with values and possibly delete the old field with the old name.

As you continue to refine the database design, you will probably make the new design available for individual projects by distributing the repository rather than the UML. Most of the time you won't need to update existing project databases with the new design, except for some special cases like typographic errors or omissions.

For an enterprise implementation with one large database, many people are reluctant to update the database using the CASE tools, since there is a chance of unexpected results from the process. Generally after a large database is in production, most groups prefer to make all of their changes using ArcCatalog or ArcSDE administration commands or both.

This chapter provides an introduction to Arc Hydro implementation. The approach may appear complex, but in practice the concepts are simple. You have significant flexibility in how you develop and deploy Arc Hydro.

A

absolute measure
Measure assigned in distance units, such as feet, meters, miles, or kilometers. *See also* relative measure.

abstract class
A class carrying attributes that are inherited by all classes below it in the class hierarchy. An abstract class does not have objects.

ancillary role
An attribute of Junction features that indicates their character: 0 = None, 1 = Source, 2 = Sink.

AreaSqKm
A descriptive attribute for area in square kilometers. This is to ensure that true areas exist for water bodies and drainage areas regardless of map units.

B

basin
A set of administratively chosen drainage areas that partition a region for purposes of water resources management. Basins are normally named after the principal rivers and streams of the region. Basins may serve as spatial packaging units for Arc Hydro data sets. *See also* catchment, watershed.

bridge
A structure occurring where a road or railroad crosses a stream.

C

catchment
A tessellation or subdivision of a basin into elementary drainage areas defined by a consistent set of physical rules. *See also* basin, watershed.

channel
A river or stream conduit or water course carrying water flow under gravity. The water-surface profile is sloping and the flow has a significant velocity, in contrast to a Waterbody where the water-surface profile is horizontal and the flow velocity is very small. *See also* Waterbody.

ChannelFeature
An abstract class carrying attributes common to ProfileLines and CrossSections.

coastline
The interface between land and water for a coastal water body. *See also* shoreline.

coded value domain
A formally defined set of integer values for an attribute where each value has an associated text descriptor. The attribute may only take on those values contained in its coded value domain.

ComplexEdge
A generic ESRI ArcObject representing a network edge that can have interior junctions.

Connector
A HydroEdge feature type used to connect short sections of the network where the line type is not known.

constructed channel
An open channel constructed in the landscape, such as a canal, ditch, or lined river channel.

CrossM
The measure location on a CrossSection (e.g., distance from the left-hand end of the CrossSection). *See also* ProfileM.

CrossSection
A transverse profile of a stream or river channel represented by a line drawn through the base of the channel perpendicular to the direction of flow. *See also* ProfileLine.

CrossSectionPoint
An object identifying the CrossM and elevation values of a point on a channel CrossSection.

CSCode
An identifier for a river CrossSection, defined based on location on a river reach.

CSOrigin	A classifier for distinguishing the type of cross-section information: 3Dtype = 0, PolylineType = 1, LineType =2 , PointType = 3, LocationType = 4. *See also* ProfOrigin.
culvert	A conduit for water flow under a roadway, usually a large-diameter pipe or a bridge opening. *See also* pipeline.

D

dam	An embankment or structure that ponds water to create a reservoir.
data model	A framework for describing a subject and storing data about it.
data set	The highest-level container of data.
DrainageArea	An area of the landscape draining to a point on a river network, to a river segment, or to a water body. Represented in the data model by an abstract class carrying attributes common to all drainage area features.
DrainageFeature	A point, line, or area feature contained within the drainage system of the landscape. Represented in the data model by an abstract class carrying attributes common to all drainage features.
DrainageLine	A line drawn through the center of the cells along a DEM-derived drainage path. Also called synthetic stream or DEM stream.
DrainagePoint	A point at the center of a digital elevation model cell, where the cell serves as the outlet cell of a DEM-derived drainage area. Also called seed point or pour point.
DrainID	An identifier for features within a particular drainage area.

E

EdgeType

A classifier for dividing HydroEdges into two types: Flowline = 1, Shoreline = 2.

EDNA

Elevation Derivatives for National Applications. A program of the U.S. Geological Survey and National Weather Service generating a catchment data set for the United States.

elevation

Elevation above mean sea level.

enabled

An attribute of network features that indicates whether the feature is traversable. Enabled = 0 means False; Enabled = 1 means True.

F

feature

A generic ESRI ArcObject representing a point, line, or area with tabular attributes. An object with spatial coordinates.

feature class

A collection of feature objects.

feature data set

A geodataset composed of graphs, networks, and feature classes.

FeatureID

HydroID of the associated hydro feature. Used to create associations between features and related information, such as time series and points on the schematic network.

FlowDir

A descriptor for flow direction on a HydroEdge. Uninitialized = 0, WithDigitized = 1, AgainstDigitized = 2, Indeterminate = 3. WithDigitized is flow in the direction of line digitizing, AgainstDigitized is flow opposite to the direction of line digitizing.

Flowline

A line tracing the path of flow through a stream, river, or water body. A subtype of HydroEdge.

FMeasure

The measure location of the beginning of a linear event. *See also* Measure, TMeasure.

FNodeID	HydroID of the SchematicNode at the "from" end of SchematicLink.
FType	A classifier for the types of hydrography features found on maps (e.g., stream/river, canal/ditch, well, swamp/marsh). Used for cartographic representation of hydrography. Can be used to create subtypes of hydrography classes.

G

geodatabase	A collection of geospatial data stored in a relational database format.
geodataset	A data set that contains geographic data.
geometric network	A topologically connected set of edges and junctions. A geometric network is a type of graph that represents a one-dimensional network, such as a utility or transportation system.
graph	Represents a set of topologically related feature classes.
GTOPO30	A 30-arc-second (1-kilometer) digital elevation model of the earth.

H

hydro network	A geometric network tracing water movement through streams, rivers, and water bodies.
HydroArea	A general class for areal hydrographic features. May be subtyped using FType.
HydroCode	The permanent public identifier of a hydro feature. The HydroCode is used to acquire information about the feature from information systems external to Arc Hydro.
HydroEdge	A complex edge feature representing a line in the hydro network. *See also* HydroEdgeType.
HydroEdgeType	The name for the coded value domain storing the EdgeType values for HydroEdges. *See also* HydroEdge.

HydroEvent   An abstract class carrying the MeasureID for point and line events.

HydroFeature   A point, line, or area describing water features of the landscape. All features of the Arc Hydro data model are hydro features described by a HydroID and a HydroCode.

HydroFlowDirections   The coded value domain containing the values for the FlowDir attribute.

Hydrography   An abstract class carrying attributes of hydrography features. In general, hydrography contains the "blue lines" on maps and point features created from tabular data inventories.

HydroID   A feature identifier populated throughout the data model that is unique within a geodatabase (e.g., HydroID = 12000164 is feature 164 in class 12).

HydroJunction   A hydro network junction that is a point of strategic hydrologic interest, such as the outlet of a water body or watershed.

HydroLine   A general class for linear hydrographic features. May be subtyped using FType.

HydroPoint   A general class for point hydrographic features. May be subtyped using FType.

HydroResponseUnit   A set of polygons having representative properties for partitioning precipitation into evaporation, infiltration, and runoff at the land surface (e.g., soil, land use, slope, administrative areas, or some combination of these with drainage boundaries).

I

inheritance   A mechanism by which objects or classes in a UML diagram acquire attributes and behaviors from objects and classes located above them in the hierarchy.

J

JunctionID

HydroID of the associated HydroJunction. Used to create associations between features and junctions on the hydro network.

L

LeftBank

A channel ProfileLine tracing the longitudinal profile of the top of the left bank of the channel, where left is taken as left to right looking in the downstream direction.

LeftFloodLine

A channel ProfileLine tracing the longitudinal profile of the flowline of water in the left floodplain or the extent of the left floodplain, where left is taken as left to right looking in the downstream direction.

LengthDown

Length along the shortest path to a downstream reference location on the hydro network.

LengthKm

A descriptor for HydroEdge length that is maintained in kilometers independent of the units of the map projection.

M

Measure

An attribute, m, attached to line vertices that measures their location along the line (e.g., river mile). *See also* FMeasure, TMeasure.

MonitoringPoint

A permanent monitoring site, such as a stream gage, rain gage, or climate station, or a sampling point on a stream or river where water-quality samples are taken periodically.

N

name	Geographic name of the feature.
natural channel	An open channel created by natural fluvial processes in the landscape.
navigation	A process of moving from one feature to the next through a landscape.
NED	National Elevation Dataset of the United States. This is a seamless, 1-arc-second (30-meter) grid of land-surface terrain for the United States.
network feature	A feature that participates in a geometric network.
NextDownID	The HydroID of the next downstream feature in a class.
NHD	National Hydrography Dataset of the United States. This is a digital version of the blue-line hydrography on USGS topographic maps.

O

object	A generic ESRI ArcObject representing a row in a data table.
object class	A collection of objects stored in a table. An object class is a type of table that stores nonspatial information.
ObjectID	A unique identifier for an object within a feature or object class. Once assigned, this ObjectID is not used again during the life of the geodatabase.
Offset	An event identifier that determines the lateral displacement of the event left or right of the line on which the event is defined. Used to avoid overlapping line events in map display or to allow events to be defined separately for left and right banks of streams or rivers.

P

pipeline	A closed conduit for water flow whose length is long relative to its diameter. *See also* culvert.
ProfileLine	A longitudinal profile of a stream or river channel. A line drawn in the direction of flow. *See also* CrossSection.
ProfileM	The measure location on a thalweg profile line. *See also* CrossM.
ProfOrigin	A classifier for distinguishing the origin of the data used to form the ProfileLine. *See also* CSOrigin.

R

reach	A length of a river or stream, usually defined between confluences. Can also refer to a water body or a portion of a water body.
reach measure	Measure defined on river reaches. Uses ReachCode as the measure identifier. *See also* river measure.
ReachCode	An identifier for a river or stream segment, usually defined from one confluence to the next. Used for linear referencing of events along rivers and streams (e.g., ReachCode = 11030406000361 identifies river segment 361 within drainage area 11030406).
relational database	A collection of related tables describing a subject. Microsoft Access, SQL Server, and Oracle are relational databases.
relationship	An association between objects in two classes using common attribute values in key fields.
relative measure	Measure defined in units of percent distance, 0–100, along a line. *See also* absolute measure.
RightBank	A channel ProfileLine tracing the longitudinal profile of the top of the right bank of the channel, where right is taken as left to right looking in the downstream direction.

RightFloodLine	A channel ProfileLine tracing the longitudinal profile of the flowline of water in the right floodplain or the extent of the right floodplain, where right is taken as left to right looking in the downstream direction.
river measure	Measure defined from a reference point on a river (e.g., river mile or kilometer). Uses RiverCode as the measure identifier. *See also* reach measure.
RiverCode	An identifier for a river, defined by the name of the river or by the concatenation of the latitude and longitude of its outlet location.
row	A record in a table. All rows in a table share the same set of fields.

S

SchematicLink	A line in a schematic network connecting hydro features.
SchematicNode	A point in a schematic network connecting hydro features.
seed point	*See* DrainagePoint.
Shape	A field in the geodatabase table for feature class that stores the geographic coordinates of the feature.
Shape_Area	The area of a shape as measured in the spatial units defined for a feature data set.
Shape_Length	The length of a shape as measured in the spatial units defined for a feature data set.
shoreline	The interface between land and water for an inland coastline water body. *See also* coastline.
Simple Junction	A generic ESRI ArcObject representing a point or a geometric network.
sink	A junction at which flow discharges from a network.
source features	Features drawn from the hydrographic data layers of a map series.
structure	Any other kind of water resources structure besides dams and bridges (e.g., buildings at risk from flooding).
synthetic stream	*See* DrainageLine.

T

table	A collection of rows that have attributes stored in columns.
thalweg	A channel ProfileLine tracing the longitudinal profile of the lowest elevation in the stream channel.
TIN	*See* triangulated irregular network.
TMeasure	The measure location of the end of a linear event. *See also* FMeasure, measure, MeasureID.
TNodeID	HydroID of the SchematicNode at the "to" end of SchematicLink.
triangulated irregular network	A mesh of triangles describing a surface, such as land terrain. Also called a TIN.
TSDateTime	The time for a particular time series value.
TSType	The type of time series (e.g., Precipitation, Streamflow).
TSValue	The time series value.

U

UserPoint	Any otherwise unclassified point of interest in the water system, such as the confluence point of two rivers, the point where a river crosses an aquifer recharge zone boundary, or the point where a river crosses an administrative boundary.

W

Waterbody	An areal water feature. A Waterbody may contain many HydroArea polygons depicting detail, such as islands within the water body. *See also* channel.
Waterbody flowline	A line through the center of a waterbody that traces the principal pathway of water movement.

WaterDischarge   A location where water is discharged to the natural water system (e.g., the discharge point of a wastewater treatment plant).

watershed   A tessellation or subdivision of a basin into drainage areas selected for a particular hydrologic purpose. Watersheds may drain to points on a river network, to river segments, or to water bodies. *See also* catchment, basin.

WaterWithdrawal   A location where water is extracted from the natural water system, usually by pumping (e.g., the intake point for a water supply).

workspace   A folder or directory in which data sets are stored.

References and resources

Chapter 1

Hydro-Climatic Data Network: water.usgs.gov/GIS/metadata/usgswrd/hcdn.html

Chapter 2

Clarke, J., et al. 2001. *Accessing the Geodatabase*. Redlands, Calif.: ESRI Press.

Hirsch, R. M. 2001. *Vision for the national geospatial framework for surface water.* Presented at the National Hydrography Dataset Applications Symposium, CRWR Online Report 01-01, Center for Research in Water Resources, University of Texas at Austin.

The David Rumsey Historical Map Collection is a private collection of historic maps, focusing on rare eighteenth- and nineteenth-century and North and South America cartographic history materials: www.davidrumsey.com.

Chapter 4

Verdin, K. L. and J. P. Verdin. 1999. A topological system for delineation and codification of the earth's river basins. *Journal of Hydrology* 218 (1-2):1-12.

Chapter 5

Cunge, J. A., F. M. Holly, Jr., and A. Verwey. 1980. *Practical Aspects of Computational River Hydraulics.* London: Pitman Publishing Limited.

Djokic, D., and D. R. Maidment. 2000. *Hydrologic and Hydraulic Modeling Support with Geographic Information Systems.* Redlands Calif.: ESRI Press.

Naiman, Robert J., and Robert E. Bilby, eds. 1998. *River Ecology and Management.* N.Y.: Springer-Verlag.

More information about HEC-RAS can be found at www.hec.usace.army.mil

Chapter 6

The U.S. Army Corps of Engineers maintains a National Inventory of Dams (www.tec.army.mil/nid) that lists thousands of dams in the country.

The U.S. Geological Survey publishes the latitude and longitude of its monitoring stations (water.usgs.gov/usa/nwis) and the U.S. Environmental Protection agency maintains a Permit Compliance System for water supply and wastewater discharge facilities (www.epa.gov/enviro/html/water.html#PCS).

The National Hydrography Dataset (NHD) is a data model for representation of map hydrography, developed by the USGS and EPA (see nhd.usgs.gov).

River Reach File 3 is presently distributed with EPA's Basins program for Total Maximum Daily Load analysis (see www.epa.gov/OST/BASINS).

A regional hydrography framework for the Pacific Northwest is available at framework.dnr.state.wa.us.

The Wisconsin Department of Natural Resources has developed detailed data-capture and feature-coding decision rules for its 1:24,000 hydrography data layer (see www.dnr.state.wi.us/org/at/et/geo/data/hyd24k.html).

Chapter 8

United States uses eight-digit Hydrologic Unit Code watersheds as data-packaging units (see www.epa.gov/ost/basins for more information about Basins).

Djokic, D., A. Coates, and J. E. Ball. 1995. GIS As Integration Tool For Hydrologic Modeling: A Need For Generic Hydrologic Data Exchange Format. ESRI User Conference, 1995.

Djokic, D. and D. R. Maidment. 2000. *Hydrologic and Hydraulic Modeling Support with Geographic Information Systems.* Redlands, Calif.: ESRI Press.

LibHydro Users Manual. 1995. U.S. Army Corp of Engineers, Hydrologic Engineering Center, Calif.

Quenzer, A. M., and D. R. Maidment. 1998. A GIS Assessment of the Total Loads and Water Quality in the Corpus Christi Bay System, CRWR Online Report, 98-1, Center for Research in Water Resources, University of Texas at Austin (www.crwr.utexas.edu/online.shtml)

Chapter 9

MacDonald, Andrew. 2001. *Building a Geodatabase.* Redlands, Calif.: ESRI Press.

Vienneau, Aleta. 2001. *Using ArcCatalog.* Redlands, Calif.: ESRI Press.

Zeiler, Michael. 1999. *Modeling Our World: The ESRI Guide to Geodatabase Design.* Redlands, Calif.: ESRI Press.

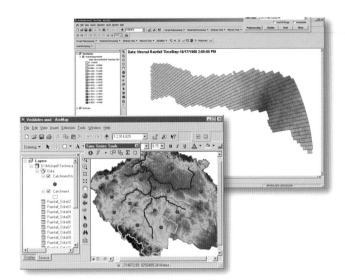

The Center for Research in Water Resources (CRWR) of the University of Texas at Austin, and the Environmental Systems Research Institute, Inc. (ESRI) have established a Consortium for developing and implementing new geographic information system (GIS) capabilities in water resources. The initial focus of the consortium is on design of a new geodatabase model for rivers and watersheds for ArcInfo version 8. The consortium is headed by Dr. David R. Maidment, director of CRWR, and supported by software development staff at ESRI. Interested organizations and individuals are invited to join the consortium.

www.crwr.utexas.edu/giswr

Advanced Spatial Analysis: The CASA Book of GIS *1-58948-073-2*
ArcGIS and the Digital City: A Hands-on Approach for Local Government *1-58948-074-0*
ArcView GIS Means Business *1-879102-51-X*
A System for Survival: GIS and Sustainable Development *1-58948-052-X*
Beyond Maps: GIS and Decision Making in Local Government *1-879102-79-X*
Cartographica Extraordinaire: The Historical Map Transformed *1-58948-044-9*
Cartographies of Disease: Maps, Mapping, and Medicine *1-58948-120-8*
Children Map the World: Selections from the Barbara Petchenik Children's World Map Competition *1-58948-125-9*
Community Geography: GIS in Action *1-58948-023-6*
Community Geography: GIS in Action Teacher's Guide *1-58948-051-1*
Confronting Catastrophe: A GIS Handbook *1-58948-040-6*
Connecting Our World: GIS Web Services *1-58948-075-9*
Conservation Geography: Case Studies in GIS, Computer Mapping, and Activism *1-58948-024-4*
Designing Better Maps: A Guide for GIS Users *1-58948-089-9*
Designing Geodatabases: Case Studies in GIS Data Modeling *1-58948-021-X*
Disaster Response: GIS for Public Safety *1-879102-88-9*
Enterprise GIS for Energy Companies *1-879102-48-X*
Extending ArcView GIS (version 3.x edition) *1-879102-05-6*
Fun with GPS *1-58948-087-2*
Getting to Know ArcGIS Desktop, Second Edition Updated for ArcGIS 9 *1-58948-083-X*
Getting to Know ArcObjects: Programming ArcGIS with VBA *1-58948-018-X*
Getting to Know ArcView GIS (version 3.x edition) *1-879102-46-3*
GIS and Land Records: The ArcGIS Parcel Data Model *1-58948-077-5*
GIS for Everyone, Third Edition *1-58948-056-2*
GIS for Health Organizations *1-879102-65-X*
GIS for Landscape Architects *1-879102-64-1*
GIS for the Urban Environment *1-58948-082-1*
GIS for Water Management in Europe *1-58948-076-7*
GIS in Public Policy: Using Geographic Information for More Effective Government *1-879102-66-8*
GIS in Schools *1-879102-85-4*
GIS in Telecommunications *1-879102-86-2*
GIS Means Business, Volume II *1-58948-033-3*
GIS Tutorial: Workbook for ArcView 9 *1-58948-127-5*
GIS, Spatial Analysis, and Modeling *1-58948-130-5*
GIS Worlds: Creating Spatial Data Infrastructures *1-58948-122-4*
Hydrologic and Hydraulic Modeling Support with Geographic Information Systems *1-879102-80-3*
Integrating GIS and the Global Positioning System *1-879102-81-1*
Making Community Connections: The Orton Family Foundation Community Mapping Program *1-58948-071-6*
Managing Natural Resources with GIS *1-879102-53-6*
Mapping Census 2000: The Geography of U.S. Diversity *1-58948-014-7*
Mapping Our World: GIS Lessons for Educators, ArcView GIS 3.x Edition *1-58948-022-8*
Mapping Our World: GIS Lessons for Educators, ArcGIS Desktop Edition *1-58948-121-6*
Mapping the Future of America's National Parks: Stewardship through Geographic Information Systems *1-58948-080-5*
Mapping the News: Case Studies in GIS and Journalism *1-58948-072-4*
Marine Geography: GIS for the Oceans and Seas *1-58948-045-7*
Measuring Up: The Business Case for GIS *1-58948-088-0*
Modeling Our World: The ESRI Guide to Geodatabase Design *1-879102-62-5*
Past Time, Past Place: GIS for History *1-58948-032-5*

Continued on next page

Books from ESRI Press (continued)

Planning Support Systems: Integrating Geographic Information Systems, Models, and Visualization Tools *1-58948-011-2*
Remote Sensing for GIS Managers *1-58948-081-3*
Salton Sea Atlas *1-58948-043-0*
Spatial Portals: Gateways to Geographic Information *1-58948-131-3*
The ESRI Guide to GIS Analysis, Volume 1: Geographic Patterns and Relationships *1-879102-06-4*
The ESRI Guide to GIS Analysis, Volume 2: Spatial Measurements and Statistics *1-58948-116-X*
Think Globally, Act Regionally: GIS and Data Visualization for Social Science and Public Policy Research *1-58948-124-0*
Thinking About GIS: Geographic Information System Planning for Managers (paperback edition) *1-58948-119-4*
Transportation GIS *1-879102-47-1*
Undersea with GIS *1-58948-016-3*
Unlocking the Census with GIS *1-58948-113-5*
Zeroing In: Geographic Information Systems at Work in the Community *1-879102-50-1*

Forthcoming titles from ESRI Press

A to Z GIS: An Illustrated Dictionary of Geographic Information Systems *1-58948-140-2*
Charting the Unknown: How Computer Mapping at Harvard Became GIS *1-58948-118-6*
GIS for Environmental Management *1-58948-142-9*
GIS for the Urban Environment *1-58948-082-1*
Mapping Global Cities: GIS Methods in Urban Analysis *1-58948-143-7*

Ask for ESRI Press titles at your local bookstore or order by calling 1-800-447-9778. You can also shop online at www.esri.com/esripress. Outside the United States, contact your local ESRI distributor.

ESRI Press titles are distributed to the trade by the following:

In North America, South America, Asia, and Australia:
Independent Publishers Group (IPG)
Telephone (United States): 1-800-888-4741 • Telephone (international): 312-337-0747
E-mail: frontdesk@ipgbook.com

In the United Kingdom, Europe, and the Middle East:
Transatlantic Publishers Group Ltd.
Telephone: 44 20 8849 8013 • Fax: 44 20 8849 5556 • E-mail: transatlantic.publishers@regusnet.com

ESRI Press • 380 New York Street • Redlands, California 92373-8100 • www.esri.com/esripress